Functional Foods

Functional Foods

Dr. Sara Diana Garduño Diaz

Delve
PUBLISHING
www.delvepublishing.com

Functional Foods

Dr. Sara Diana Garduño Diaz

Delve Publishing
2010 Winston Park Drive,
2nd Floor
Oakville, ON L6H 5R7
Canada
www.delvepublishing.com
Tel: 001-289-291-7705
001-905-616-2116
Fax: 001-289-291-7601
Email: orders@arclereducation.com

ISBN: 978-1-77361-278-2 (Hardcover)

ABOUT THE AUTHOR

Dr. SARA DIANA GARDUÑO DIAZ

Dr. Sara D. Garduño-Diaz has a background in nutrition with a PhD in Nutrition and Food Science from the University of Leeds, United Kingdom; she has further specialized in Sports Nutrition with the International Olympics Committee, Switzerland. Dr Sara has worked as Research Fellow for several EU-funded research projects and was awarded research grants from CONACYT (Mexico) and the BBSRC (UK). She is currently Senior Nutrition Consultant at Your Choice Nutrition in Kuwait, where she has been working on functional nutrition and wellness since 2014.

Dr Sara sits on the editorial board of various international journals, including the Journal of Obesity and Weight Management, E Cronicon Nutrition, Research in Health Science, Insights in Nutrition and Dietetics and Nourish, for whom she is the Executive Editor. She is the author of several peer-reviewed publications and a book: Diet, Ethnicity and the Metabolic Syndrome (Saarbrucken, Germany: 2013). Her research interests include the dietary patterns of migrant populations and their impact on health, as well as investigating the environmental factors that influence food selection.

Dr. Garduño-Diaz is a member of the Academy of Nutrition and Dietetics, the American Overseas Dietetics Association, Professionals in Nutrition for Exercise & Sport, and the World Public Health Nutrition Association.

With over 10 years of experience in the field of nutrition, and having lived and worked in various continents, Dr Sara's approach to food is one of integrating strategies to design individual programs based on her client's needs. Dr Sara advocates for real food and the art of eating.

TABLE OF CONTENTS

LIST OF ABBREVIATIONS

ACE	Angiotensin I-Converting Enzyme
ADA	American Dietetic Association
ALA	Alpha-Linolenic Acid
AMD	Age-related Macular Degeneration
AODF	Anti-Oxidant Dietary Fibers
BMD	Bone Mineral Density
BPA	Bisphenol A
CHD	Coronary Heart Disease
CLA	Conjugated linoleic acid
CVD	Cardiovascular Disease
DF	Dietary Fibres
DHA	Docosahexaenoic Acid
EFSA	European Food Safety Authority
EPA	Eicosapentaenoic Acid
EPIC	European Prospective Investigation into Cancer and Nutrition
FAO	Food and Agriculture Organization
FDA	Food and Drug Administration
FIM	Foundation for Innovation in Medicine
FOSHU	Foods for Specified Health Use
GI	Gastro-Intestinal
GM	Genetically modified
GRAS	Generally Recognized as Safe
HDL	High Density Lipoprotein
HPFS	Health Professionals Follow-up Study
IBS	Irritable Bowel Syndrome
IFIC	International Food Information Council
IFT	Institute of Food Technologists
ILSI	International Life Sciences Institute of North America
ISP	Isolated Soy Protein

LDL	Low Density Lipoprotein
MRP	Maillard reaction products
MUFA	Mono-Unsaturated Fatty Acids
NCCIH	National Centre for Complementary and Integrative Health
NHIS	National Health Interview Survey
NHS	Nurses' Health Study
NSAIDS	Nonsteroidal Anti-Inflammatory Drugs
OSC	Organosulfur Compounds
PUFA	Poly-Unsaturated Fatty Acids
SCFA	Short Chain Fatty Acids
SIBO	Small Intestine Bacterial Overgrowth
TBARS	ThioBarbituric Acid Substances
TG	Triglycerides
USDA	United States Department of Agriculture
WHO	World Health Organization

LIST OF FIGURES

LIST OF TABLES

PREFACE

Clearly, all foods are functional, as they provide some degree of flavour, scent, or nutrient value. Within the last years, however, the term functional as it applies to food has adopted a different implication: that of providing an additional physiological benefit beyond that of meeting basic nutritional needs of energy, vitamins or minerals. Enter functional foods.

As consumers worldwide become more health conscious, the demand for health-promoting foods and food components is expected to grow. The market for such foods is predicted to be quite large. Before the full market potential can be realized, however, consumers will need to be assured of the safety and efficacy of functional foods. Current and future scientific studies are expected to provide this assurance and to inspire confidence in functional foods in the minds of consumers worldwide. As the products of scientific innovation and industrial processing, functionally enhanced foods promise healthfulness but conflict with understanding held by many consumers that healthy foods are those that are the most "natural."

This book explores these conflicting ideas: that of developing new food products, and that of keeping them familiar enough to ensure acceptance and consumption. The book covers definition of key concepts, such as nutraceuticals and bioavailability. It explores the different types of functional foods, both naturally occurring and "designer foods." Sections on key chemical components are included, as are discussions on how these compounds interact as part of different foods. Finally, sections on what functional foods represent for the consumer, health and safety issues, as well as legislation and potential risks with functional foods are covered.

Several challenges exist for the vitality of the functional foods market in the years ahead. First, there are a number of scientific challenges for those attempting to unlock the food/health "equation." How do these food components act to modify disease processes and for whom are these benefits most pronounced? The food/health equation includes a complex set of interactions that must be better understood so that new food product efficacy can be appropriately characterized. The factors contributing to disparities in individual health responses to foods need to be clarified. These include genetic variation, gender differences, individual differences in metabolism, and how bioactive phytochemicals alter the routine functions of traditional nutrients.

This is a book of exploration, designed for undergraduate and graduate students with a basic understanding of sciences. With a well-rounded approach, it aims to raise points for consideration and discussion, and to challenge both supporters and antagonists of the role of functional food in health.

Author

SYNOPSYS

Functional foods can be considered to be those whole, fortified, enriched or enhanced foods that provide a health benefit beyond the provision of essential nutrients (including energy, vitamins and minerals), when they are consumed regularly at sufficient levels as part of a varied diet. Linking the consumption of functional foods or functional food ingredients with health claims should be based on sound scientific evidence, ideally being replicated through randomized, placebo-controlled, intervention trials in human subjects; and if possible with observational studies of real life scenarios. Still, not all foods available on the market today that claim to be a functional food are supported by sufficient sound evidence to justify such claims. This book explores the concept of functional foods, their possible mechanism of action, the available evidence supporting their functionality role, the strength of that evidence and the recommended uses of such products. Concepts regarding customer consumption, marketing, awareness and safety are also touched upon. Finally, the functional food regulations including identification and labelling are discussed as well as current trends and future potential for development of the field of functional foods. Functional foods represent one of the most exciting areas of development in the food and nutrition sciences today. However, it must be emphasized that these foods and ingredients are not magic, nor are they substitutes for poor health habits. Favourable eating habits, with the inclusion of true functional foods, are only one aspect of a comprehensive approach to wellness.

FUNCTIONAL FOODS

INTRODUCTION

A historical perspective

If we consider that all foods provide a certain flavour, smell and nutrients; then it is possible to consider them all as functional. Nevertheless, in more recent times, the term "functional foods" has come to refer to any food or food product that can deliver an additional physiological benefit beyond just the provision of energy and other nutrients.

The idea that food may be used for medicinal purposes is not something new. The powers of food were however pushed aside during the 19th century with the advances of modern medicine. Much later, in the 1900s, the vital role of diet in disease prevention and health promotion came to the forefront once again. Today, with the wellness movement in full force, it is possible to find not only practitioners and enthusiasts who are looking for the latest development in functional food; these products are also being demanded by the lay consumer.

During the first half of the 1900s scientists were focused on the identification and isolation of "vital" elements. These later came to be known as vitamins. Interest was high on understanding their role on health and any possible conditions that may arise from deficitsthereof.

Undernutrition continues to play a key research interest area until the 1970, when attention shifted to conditions related to "overnutrition." This then became the focus of public health. Guidelines then began to pop up, with official national and international recommendations as to what and how people should be nourishing their bodies. These included the *Dietary Guidelines for Americans (1980, 1985, 1990, 1996, 2000*— a joint publication of the USDA and the Department of Health and Human Services), the *Surgeon General's Report on Nutrition and Health* (1988), the National Research Council's *Diet and Health* (1989) and *Healthy People 2000* and *2010* from the U.S. Public Health Service; and these are only the official guidelines in the United States alone!

Public health was in full force, with interest on food and its components shifting from scientists only, to policy makers, health professionals and demanding consumers world-wide. The importance of consuming a diet containing beneficial foods was recognized and everyone wanted to know more. The aim of course was not only to avoid disease, but to begin focusing on reaching optimal health.

In tandem, scientists began to recognize physiologically active components in both animal and plant-based foods. Today these are known as phytochemicals and zoochemical, respectively. These scientific developments, joined with an aging, health-conscious population, changes in food regulations, technological progress and a marketready for the introduction of new health-promoting products, all came together in the 1990s to create the trend we currently known as "functional foods."

What are Functional Foods

The term "functional food" is strongly used and often exploited by the giants of the food industry. Although not recognized officially at an international level, numerous working definitions have been proposed. These include the definition given by Health Canada, where:

A functional food is 'similar in appearance to conventional food, consumed as part of the usual diet, with demonstrated physiological benefits, and/or to reduce the risk of chronic disease beyond basic nutritional functions' (Health Canada, 2000).

There is also the definition from the International Food Information Council (IFIC), which states:

> Functional foods are 'foods or dietary components that may provide a health benefit beyond basic nutrition' (Bagchi, 2008).

The International Life Sciences Institute of North America (ILSI) in turn defines functional foods as:

> 'foods that by virtue of physiologically active food components provide health benefits beyond basic nutrition' (Bagchi, 2008).

While the European Commission Concerted Action on Functional Food Science in Europe regards a food to be functional:

> 'if it is satisfactorily demonstrated to affect beneficially one or more target functions in the body, beyond adequate nutritional effects, in a way that is relevant to either an improved state of health and well-being and/or reduction of risk of disease' (Consensus document, 1999).

For purposes of this book, a functional food is defined as:

> Any food that exerts health properties beyond the traditional nutrients it contains.

With this definition in mind, two categories of functional foods will be covered:

1. Foods that naturally contain biologically active, non-nutrient compounds that provide health benefits.
2. Food products specifically formulated to have higher amounts of nutrients than would naturally occur in that food.

Health-aware and highly-informed consumers are increasingly seeking functional foods in an effort to manage their own health and well-being.However, foods are now being examined intensively for added physiological benefits, which may reduce chronic disease risk or otherwise optimize health. It is these research efforts that have led to the global interest in the growing food category of functional foods.

Table 1: What can be considered a functional food?

A functional food can be:
(1) a natural food
(2) a food to which a component has been added
(3) a food from which a component has been removed
(4) a food where one or more components has been modified
(5) a food in which the bioavailability has been modified
(6) any combination of these.

Attribution has been given to the Japanese for first establishing the concept of functional foods. This has been traced back to the 1980s, when, encountering ever increasing costs for the health care of their population, The Ministry of Health and Welfare set up an official system to regulate the establishment of certain foods as beneficial for improving the health of their aging population. (Arai, 1996). It was required that these benefits be well documented and that any health claims be proved before official recognition of a food or food product as functional could be established by the Japanese government.

In the United States, functional foods have no such regulatory identity as they do in Japan. For the time being, expert food scientists are working on establishing an official definition for functional foods with several organizations having proposed definitions for this relatively new food category.

In 1994, the National Academy of Sciences' Food and Nutrition Board defined functional foods as

> "any modified food or food ingredient that may provide a health benefit beyond the traditional nutrients it contains" (CON, 1994).

In 1999 the American Dietetic Association defined functional foods as foods that are:

"whole, fortified, enriched, or enhanced," but more importantly, states that such foods must be consumed as "… part of a varied diet on a regular basis, at effective levels" for consumers to reap their potential health benefits (ADA, 1999).

In addition to government agencies, national and international organizations developed their own definitions for functional food:

The Institute of Food Technologists (IFT) defines "functional foods" as:

"Substances that provide essential nutrients often beyond quantities necessary for normal maintenance, growth, and development, and/or other biologically active components that impart health benefits or desirable physiological effects" (Clydesdale, 2004).

This definition is comparable to the EU definition in that they highlight the idea that functional food improves health in addition to providing basic nutrition.

In the U.S. alone, scientific organizations interpret the term "functional food" differently. Although IFT, The National Academy of Sciences, and the ADA concur on the essence of what functional food is, their definitions vary in detail, specificity, and focus. This fact demonstrates the need for a unifying definition in the United States and abroad. After a formal definition is established, scientific organizations will more easily be able discuss about the future of functional food.

Common themes in the definition of functional foods

The functional food definitions proposed by various organizations have common themes:

- Functional foods are part of the usual diet.

Health Canada, DC, the Academy of Nutrition and Dietetics, and FAO describe functional foods as being part of the usual or regular diet (Health Canada, 19998; FAO, 2007; ADA, 2009; Dietitians of Canada, 2010). According to the Academy of Nutrition and Dietetics, functional foods must be consumed on a regular basis in order to be effective, and that consuming a functional food only once is unlikely to provide the associated health benefit (ADA, 2009).

- Functional foods are in food forms.

Each definition specifies that functional foods exist in food forms. Health Canada and FAO provide an easy rule of thumb by stating that functional foods are 'similar in appearance' to conventional foods (Health Canada, 19998; FAO, 2007.

- Functional foods contain an ingredient or bioactive which provides a health benefit.

Health Canada, DC and FAO specify that functional foods, and the bioactives they contain, provide health benefits beyond basic nutritional needs (Health Canada, 19998; FAO, 2007; ADA, 2009) This is reflective of the evolution in nutrition research over the past few decades (FAO, 2007; ADA, 2009). The focus has switched from identifying essential nutrients to treat deficiency diseases to investigating the potential of bioactives to reduce chronic disease risk, manage disease symptoms and improve overall well-being (Health Canada, 19998; FAO, 2007; ADA, 2009).

The definition work remains under progress and will then likely proceed to establishing criteria and regulations for the identification of and labeling these functional foods. The aim is then usually to translate relevant scientific findings into applicable practices for the consumer. While a definition of functional foods is paramount, it is only the first step in a chain of requirements to meet this aim.

The making of functional foods

Functional foods may be created by enhancing or fortifying an existing natural food or synthetic food product with a specific nutrient to upgrade the food as having an additional health-related benefit. Over the past five years, modified foods have increased in popularity, gaining a larger percentage of the market. Calcium-, antioxidant-, vitamin-, and herb-fortified beverages; products containing prebiotics and probiotics; and products containing added plant stanols/sterols, fibers, and omega-3 fatty acids are examples of how foods are commonly modified to increase their nutritive value. A food product can be made functional by using any of the five approaches listed below:

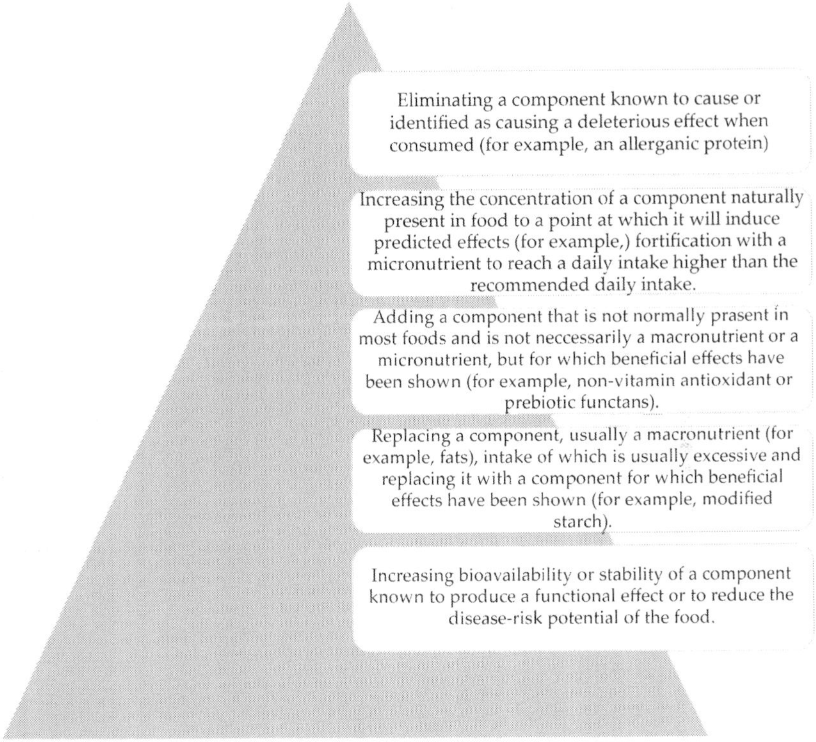

Figure 1: Means by which a food product may be made functional.

Comparable terms to "Functional Food"

As the science and technology to develop new functional foods grows, so have the number of related terms and vocabulary.

- "Nutraceutical" is often interchanged with "functional food" in scientific literature.
- "Medical foods" are used to treat metabolic problems and diseases.
- "Dietary supplements" are meant to provide extra health benefits outside of basic nutrition.

As evidenced by the number of existing nutritional terms, nutritional science is expanding. However, the similarity of its terms can obstruct the implementation of meaningful regulatory policies. Without a formal definition, functional food is lost in the wide range of food-related vocabulary. Thus, expert bodies, regulatory bodies, and most importantly, the consumers, cannot differentiate functional food from other food products.

Nutraceuticals

A term often used by the food industry interchangeably with functional foods (although it is less favored by consumers) is "nutraceuticals." This term was first coined in 1991 by the Foundation for Innovation in Medicine to refer to nearly any bioactive component that delivers a health benefit. In a 1999 policy paper, Zeisel (1999) distinguished whole foods from the isolated components derived from them in his following definition of nutraceuticals:

> "those diet supplements that deliver a concentrated form of a presumed bioactive agent from a food, presented in a nonfood matrix, and used to enhance health in dosages that exceed those that could be obtained from normal food."

The term "nutraceutical" is a combination of the terms "nutrition" and "pharmaceutical", and is thought to have been proposed by Stephen DeFelice, MD, founder and chairman of the Foundation for Innovation in Medicine (FIM), Cranford, NJ in 1989 (Brower, 19998). Nutraceuticals is a broad umbrella term that is used to describe any product derived from food sources with extra health benefits in addition to the basic nutritional value found in foods. They can be considered non-specific biological therapies used to promote general well-being, control symptoms and prevent malignant processes. DeFelice took further steps to officially define nutraceutical as:

> "a food (or part of a food) that provides medical or health benefits, including the prevention and/or treatment of a disease" (Brower, 1998).

It is important to highlight that the term nutraceutical, as commonly used in marketing and by the food industry, has no regulatory definition in most countries (Zeisel, 1999).

Nutraceuticals further differ from dietary supplements in the following aspects:

(1) nutraceuticals must not only supplement the diet but should also aid in the prevention and/or treatment of disease and/or disorder.

(2) nutraceuticals are used as conventional foods or as sole items of a meal or diet (Kalra, 2003).

With these two aspects in mind it is easy to understand how the terms "nutraceutical" and "functional food" can be easily confused and often used interchangeably. A functional food for one consumer can act as a nutraceutical for another consumer.

Examples of nutraceuticals include(Kalra, 2003):

- Minerals, vitamins and other dietary supplements
- Herbal products: garlic (allicin), ginger, echinacea, ginseng, liquorice, onion, senna, turmeric (curcumin)
- Dietary enzymes: bromelain, papain
- Dietary fibre
- Hydrolysed proteins
- Phytonutrients: resveratrol
- Carotenoids: lycopene
- Prebiotics
- Probiotics
- Fortified dairy products (e.g., milk) and citrus fruits (e.g., orange juice)

Medicinal foods

The term medicinal food, as defined by the FDA in section 5(b) of the Orphan Drug Act (21 U.S.C. 360ee (b) (3)) is:

> "a food which is formulated to be consumed or administered enterally under the supervision of a physician and which is intended for the specific dietary management of a disease or condition for which distinctive nutritional requirements, based on recognized scientific principles, are established by medical evaluation."

Medicinal foods are distinguished from the wider category of foods for special dietary use by the requirement that medicinal foods be *intended to meet distinctive nutritional requirements of a disease or condition*, used under medical supervision, and intended for the specific dietary management of a disease or condition. Medicinal foods are not those simply recommended by a physician as part of an overall diet to manage the symptoms or reduce the risk of a disease or condition. Not all foods fed to patients with a disease, including diseases that require dietary management, are medical foods. Instead, medicinal foods are foods that are specially formulated and processed (as opposed to a naturally occurring foodstuff used in a natural state) for a patient who requires use of the product as a major component of a disease or condition's specific dietary management.

Dietary supplements

A dietary supplement is a product intended for ingestion and it is one that contains a dietary ingredient intended to add further nutritional value to supplement the diet in areas that may be deficient or may be presenting increased nutrient needs. In turn, a dietary ingredient may be one, or any combination, of the following substances:

- a vitamin
- a mineral
- an herb or another botanical
- an amino acid
- a dietary substance for use by people to supplement the diet by increasing the total dietary intake
- a concentrate, metabolite, constituent, or extract

Dietary supplements may be found in many forms such as tablets, capsules, soft gels, gel caps, liquids, or powders. Some dietary supplements can help ensure that you get an adequate dietary intake of essential nutrients; others may help reduce the risk of disease.

Dietary supplements must be labelled as such and must not be represented for use as a conventional food or as the sole item of a meal or the diet. Conventional foods are foods that are not dietary supplements and that rather are the main components of the daily diet. One way to distinguish dietary supplements from conventional foods is by looking at the nutrition information on the label of the product. Conventional foods must have a "Nutrition Facts" panel on their labels, but dietary supplements must have a "Supplement Facts" panel.

Most adults in the United States take one or more dietary supplements either every day or occasionally. Popular supplements include vitamins D and E; minerals like calcium and iron; herbs such as echinacea and garlic; and specialty products like glucosamine, probiotics, and fish oils. When a person's diet may not be providing a nutritious variety of foods, some supplements might help achieve adequate amounts of essential nutrients. However, supplements should not take the place of the variety of foods that are important to a healthy diet.

Many supplements contain active ingredients that can have strong effects in the body with the possibility of unexpected side effects. Supplements are most likely to cause side effects or harm when people take them instead of prescribed medicines or when people take many supplements in combination. Some supplements can increase the risk of bleeding or, if a person takes them before or after surgery, they can affect the person's response to anesthesia. Dietary supplements can also interact with certain prescription drugs in ways that might cause problems, for example:

- Vitamin K can reduce the ability of blood thinners to prevent blood from clotting.
- St. John's wort can speed the breakdown of many drugs (including antidepressants and birth control pills) and thereby reduce these drugs' effectiveness.
- Antioxidant supplements, like vitamins C and E, might reduce the effectiveness of some types of cancer chemotherapy.

In addition, some ingredients in dietary supplements are added to a growing number of foods, including breakfast cereals and beverages. Thus, people may be getting more of these ingredients than they think, and may even be reaching critical levels and overdosing. Taking more than you need is always more expensive and can also raise your risk of experiencing side effects. For example, getting too much vitamin A can cause headaches and liver damage, reduce bone strength, and cause birth defects. Excess iron causes nausea and vomiting and may damage the liver and other organs.

Research into functional foods

Several factors are responsible for the fact that this is one of the most active areas of research in the nutrition sciences today:

1) An emphasis in nutritional and medical research on associations between diet and dietary constituents and health benefits

2) A favorable regulatory environment

3) The consumer self-care phenomenon

4) Rapid growth in the market for health and wellness products

Table 2: Working definitions of the term "functional food"

Organization	Definition
Academy of Nutrition And Dietetics	"Foods defined as whole foods along with fortified, enriched, or enhanced foods that have a potentially beneficial effect on health when consumed as part of a varied diet on a regular basis at effective levels."
International Food Information Council	"Foods or dietary components that may provide a health benefit beyond basic nutrition and may play a role in reducing or minimizing the risk of certain diseases and other health conditions."
Institute of Food Technologists	"Foods and food components that provide a health benefit beyond basic nutrition (for the intended population)."
International Life	"Foods that by virtue of the presence of physiologically active food components provide health benefits beyond basic nutrition."
European Commission	"A food that beneficially affects one or more target functions in the body, beyond adequate nutritional effects, in a way that is relevant to either an improved state of health and well-being and/or reduction of risk of disease. It is part of a normal food pattern, it is not a pill, a capsule or any form of dietary supplement."
Health Canada	"A functional food is similar in appearance to, or may, be a conventional food, is consumed as part of a usual diet and is demonstrated to have physiological benefits and/or reduce the risk of chronic disease beyond basic nutritional functions."
Japanese Ministry of Health, Labour and And Welfare	"FOSHU [food for specified health users] refers to foods containing ingredient with functions for health and officially approved to claim its physiological effects on the human body. FOSHU is intended to be consumed for the maintenance/promotion of health or special health uses by people who wish to control health conditions, including blood pressure or blood cholesterol."

Functional Foods in the Marketplace

Establishing a final definition for functional food not only brings about consensus in scientific and governmental communities; a new improved definition will pave the way for functional foods to be formally and legitimately introduced to food markets that may up till now be hesitant. By 1997, Japan, the United States, and Europe each generated $3 billion in sales with a projected global sale record of $130 billion by 2015 (Young, 1996; Heasman, 1997; Lee and Foo, 2014). However, food industrialists make claims based on differing definitions, and so, are not making scientific research, food safety, and health claim legitimacy top priorities. Functional food scientists would like to revise this process by establishing a new definition for functional food. This will allow food industrialists to base their health claims on supported research. With support from the scientific and governmental communities, bringing functional foods to markets will help billions suffering from chronic illnesses and general health problems. Below (Figure 2) is a cycle of steps that the Functional Food Centre of the USA believes will help bring functional foods to market (Danik, 2015).

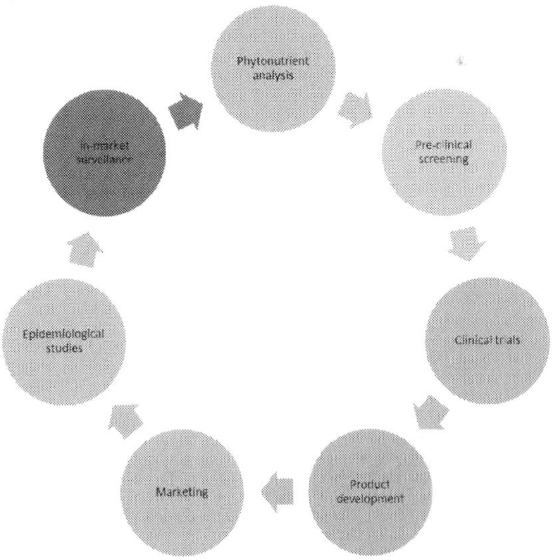

Adapted fromMartirosyan and Singh, 2015.

Figure 2: Process of introducing functional foods to the market.

Previously, foods were often fortified with nutrients to help prevent specific nutritional deficiencies. Technically, any fortified food could be considered a functional food. Some examples of commonly fortified foods include:

- iron-fortified cereals
- vitamin D-fortified milk
- iodized salt.

Today, many functional foods are aimed at boosting intakes of phytochemicals to reduce the risk for chronic diseases like cancer and heart disease. The rising prevalence of chronic diseases is prompting consumers to shift towards the consumption of healthier foods. A rise in consumer interest and an increased understanding of how a favourable diet may enhance immunity are some of the key factors driving the growth of the global functional foods market. The geriatric population is believed to benefit most from functional foods among adult consumers due to age-related problems pertaining to health. For that reason, the rising global geriatric population is boosting the growth of the functional foods market. Moreover, the rising disposable income and increasing demand for fortified food and beverage products are leading to increased uptake of functional foods, thereby propelling the growth of the market.

Although the functional foods product sales have and will continue to rise at a hurried rate, the growth of the market is hampered by various issues. The high prices of most functional food products are limiting their adoption. In addition, the global economic downturn is impelling consumers to switch to cheaper groceries that may not always meet their nutritional needs but provide sufficient energy and bulk for satiety. Furthermore, research into consumer perceptions of these products and their associated health claims is limited. Another factor restraining the demand for functional foods are food safety concerns and regulations. These factors, collectively, are impeding the growth of the market for functional foods.

Nevertheless, sustained population rise will lead to a shift of focus from high consumption areas such as the U.S. and Japan to high-population regions such as China and India. This shift is likely to offer immense growth opportunities for the functional foods market in near future. Examples of the fastest-growing functional food categories include:

- Cereals and breads with added isoflavones
- Fruit juices with herbs that have alleged immune-enhancing properties, such as echinacea
- Margarine with added phytosterols, which help to reduce cholesterol

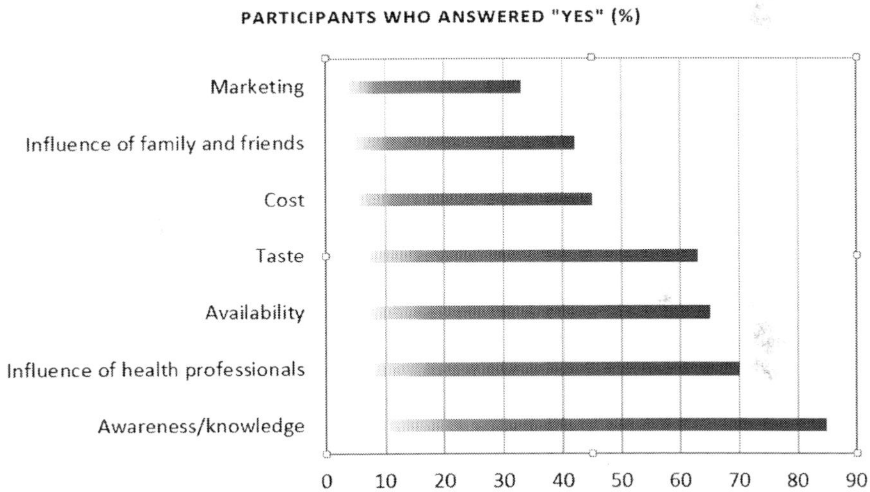

Adapted from Vella et al., 2014

Figure 3: Factors likely to increase functional food consumption.

Table 3: Advantages and disadvantages identified by users and non-users of functional foods

	User's % Agreed	Non-users % Agreed
Advantages		
Easy way to improve health	73	53
Diseases may be prevented	37	27
Less need of medicines	33	12
Medicines give you the idea you are sick, while foods Do not	9	9
It does not hurt to try	18	21
You may feel better	14	17
Uncertain about whether you ear adequately	11	8
Other	1	2
Disadvantages		
Variety in diet is preferable	20	26
Healthy foods will become very expensive	15	9
Probably it will not influence health sufficiently	3	5
Unrestricted use may be dangerous	2	4
Side effects may occur	1	3
Foods are starting to appear like a medicine	1	3
It is purely commerce	4	8
Other	-	2

Abbreviation: NS, not significant. Subjects were allowed to choose more than one reason.

Adapeted from de Jong et al., 2007

Advantages of Functional Foods

Using food as medicine or preventative medicine is part of many traditional cultures, stemming from the widespread use of ancient traditional medicine, and added to the ever-growing world population. There will be greater focus on the health benefits of food as medicine with the shift from a preference of synthetic products to more natural options. The current fast-paced lifestyle is becoming busier and more stressful than ever, resulting in more health concerns, hence the opportunity for growth of the functional food market, with functional foods making it easier to meet requirements for nutrients that are often deficient in regular diets. Some functional foods make it possible to achieve higher intakes of nutrients or phytochemicals than would be realistic with natural food sources of these compounds.

Consumer interest in the relationship between diet and health has increased substantially in Europe and other regions of the World. There is much greater recognition today that people can help themselves and their families to reduce the risk of illness and disease and to maintain their state of health and wellbeing through a healthy lifestyle, including food selection. Ongoing support for the important role of foods such as fruits and vegetables and wholegrain cereals in disease prevention and the latest research on dietary antioxidants and combinations of protective substances in plants has helped to provide the motivation for further developments in the functional food market in Europe.

Trends in population demographics and socio-economic changes also point to the need for foods with added health benefits. An increase in life expectancy, resulting in an increase in the number of elderly and the desire for an improved quality of life, as well as increasing costs of health care, have stimulated governments, researchers, health professionals and the food industry to see how these changes can be managed more effectively. There is already a wide range of foods available to today's consumer but now the impetus is to identify those functional foods that have the potential to improve health and well-being, reduce the risk from, or delay the onset of, major diseases such as cardiovascular disease (CVD), cancer and osteoporosis. Combined with a healthy lifestyle, functional foods can make a positive contribution to health and wellbeing.

Possible Disadvantages of Functional Foods:

Several problems may be highlighted with the development and consumption of functional foods:

- ***False health claims*** –Promises of effects beneficial to the health and wellbeing of the consumer and the lack of government regulation for such claims has given functional food companies free rein to make dubious allegations on behalf of their new wonder products. Often, statements are made that functional foods will perform in a certain way or produce a certain outcome, without any robust evidence that this is the case.

- ***Misleading overall diet*** –with the increased availability of functional foods there is also an increased risk that these new products might mislead consumers into switching from a diverse, healthy diet to a basically unhealthy one – with an increasing reliance on 'functional' additives or supplementation. Regardless of any potential benefits, functional foods should not fully substitute a healthy diet.
- ***Deviation of valuable research resources*** -- Much financial support that could be put into true sustainable development – promoting locally appropriate and ecologically benign agriculture is poured into the development of more potentially profitable functional food development. For example, hundreds of millions of pounds have been poured into research for 'golden rice' alone, and much more will be needed before this crop stands a chance of becoming widely available and truly beneficial to the targeted consumer population.
- ***Accessibility*** -- Most functional foods will not be accessible for those who would truly benefit from their consumption, due to the fact that they will be too expensive and inaccessible. Big Food is motivated by profits, and by the monopoly control conferred by the patents system – not by charity or scientific development. Having spent vast sums in research, even progressive companies can ill-afford to give their new products away.
- ***Power to big companies*** –With the development of novel food products, farmers are being persuaded to turn to new technologies and 'wonder crops' and throw away centuries of experience leading to further dependency on a few large corporations. Instead of relying on traditional modes of diverse and self-sufficient production, this can become a tool or level from foreign policy and control of many by a few.
- ***Technical fixes avoid the real problem*** –In some cases, the development of functional foods may become a technical fix serving as a distraction to divert attention from real [political

or economic] problems of malnourished people often at risk of poverty. Coupled with poverty, injustice, a lack of land and other resources, these problems cannot be solved by the development of functional foods alone.

- ***Exacerbates problems from the Green Revolution*** –The development of functional foods should promote growing biodiversity and not monocultures. All over the World, switching from diversity to monoculture has turned fruitful farms into wildlife deserts, poisoning water supplies and destroying wildlife. GM crops – with their emphasis on global supply lines and giant corporations – destroy local capacities to develop food security and intensify the concentration of land ownership, as the Green Revolution did before.

In addition, mmanufactured functional foods cannot duplicate all the benefits of whole plant foods, some of which are not yet known. For example, plant foods like leafy green vegetables, which are naturally rich in calcium, also provide vitamin K and vitamin C, both of which are needed for healthy bones, as well as numerous nutrients and phytochemicals that protect against chronic disease. A diet based on whole plant foods is likely to be more healthful than one based on functional foods. Functional foods may provide benefits in health terms, but should not be seen as an alternative to a varied and balanced diet and a healthy lifestyle.

Imported functional products are often 2-3 times more expensive than domestic products for consumers. In addition, consumers are more trusting of domestic/natural products, than imported products, and are very conscious of where their food is produced. Further, the certification process for functional foods is time-consuming, especially if the functional food is not already considered "generic" by the local food and drug administration body. Functional foods that contain added nutrients or phytochemicals may send confusing messages about how to plan healthful diets. Calcium-fortified orange juice is an improvement on a food that is already healthful. Calcium-fortified candy or fibre-enriched white bread, on the other hand, involve addition of beneficial substances to foods that are otherwise devoid of healthful properties.

Table 4: Key points discussed in this book

Key points:
• Functional foods are not a new concept but are however becoming more and more popular among lay consumers. • A consensus of the definition and established characteristics of functional foods is lacking. • Functional foods deliver additional or enhanced benefits over and above their basic nutritional value. • The term 'functional foods' can be viewed as encompassing a very broad range of products. • Functional foods may be naturally occurring or synthetically made. • Some functional foods are generated around a particular functional ingredient, for example foods containing probiotics, prebiotics, or plant stanols and sterols. • Other functional foods or drinks can be foods fortified with a nutrient that would not usually be present to any great extent (e.g. folic acid fortified bread or breakfast cereals). • Functional foods and drinks may provide benefits in health terms, but should not be seen as an alternative to a varied and balanced diet and a healthy lifestyle.

THE MICROBIOME

Eating habits and trends in food production and consumption have health, environmental and social impacts. Diet has implications on gut health. Gut complications, such as ulcerative colitis, Crohn's disease, irritable bowel syndrome, and gluten therapy resistant celiac, result from overgrowth and imbalance of intestinal microbial flora, and are related to the regular components of the everyday diet. While the gastrointestinal tract is relatively sterile at birth; intestinal microbiota develops after birth, with the rate and diversity of microbiotic colonization varying according to factors such as the form of birth (vaginal vs caesarean), infant nutrition (formula fed vs breastfed), antibiotic use, diet and age. To a great extent, gut health determines an individual's overall health, with more and more connections between the gut and other organs being discovered. The human gut has the following functions:

(a) breaking food down to nutrients
(b) facilitating absorption of nutrients into the blood through intestinal walls
(c) preventing foreign and toxic molecules from entering the bloodstream.

Gut failure, therefore, has a direct negative impact on human health.

Humans are a walking eco system

Within the human body, there are thousands of microbes populating the various organs. This microbiota can be found inside and outside the organism, ranging from the mouths to the gut and everywhere in between. While the clear majority of people have allotted a negative connotation to bacteria in general, and regard it as an unwanted phenomenon, scientists are leaning toward educating the lay public and changing this perception around. Bacteria are not only beneficial but also necessary for humans to thrive and reach optimal health.

The organisms that make up the human microbiota interact with each other and in communities; as well as with their surroundings (Robinson, 2010). The human microbiota shifts its population in response to environmental changes, much as the various organisms on Earth's ecosystems do. From the time it begins to acquire its unique collection of bacteria and other microorganisms, and how it is shaped throughout life, the state of the human microbiome will greatly influence the level of health enjoyed by a population and an individual, including the diseases that may be experienced, the strength of the immune system, and even the degree of weight management and mood changes experienced.

Most health care professionals have long viewed microbes as things to be gotten rid of. Viruses and bacteria, after all, are responsible for many existing health conditions. Nevertheless, if we consider the numberof existing microbes, most do not cause sickness nor are they a serious cause for concern. On the contrary, many are important for health. As researchers learn more about the microbes that help maintain health, an understanding is being reached on how subtle imbalances in the composition of the microbiome can also cause disease—and how restoring the balance may lead to potential management or cures. This new understanding may lead to more focused and effective modes of treatment. Unlike modern antibiotics, which kill beneficial microbes along with the harmful ones, new drugs may kill only harmful bacteria while leaving the friendly ones alone. Others may nurture friendly bacteria, helping them outcompete the harmful ones.

Defining the human microbiome

Defining the human microbiome has been complicated by confusion about terminology: for example, "microbiota" (the microbial taxa associated with humans) and "microbiome" (the catalogue of these microbes and their genes) are often used interchangeably. More fundamentally, however, new findings are leading to question the concepts that are central to establishing the definition of the human microbiome, such as the stability of an individual's microbiome and whether a person has one microbiome or many.

It is generally accepted that the human microbiota consists of the 10-100 trillion symbiotic microbial cells harboured by each person, primarily bacteria in the gut; the human microbiome consists additionally of the genes harboured by these cells (Turnbaugh et al., 2007). Microbiome projects around the world have been launched with the goal of understanding the roles that these symbionts play and their impacts on human health (Peterson et al., 2009 and Qin et al., 2010).

Acquiring our microbiome

Up until birth, a human is relatively sterile—with little to no microbes. From the moment of birth microbes begin the colonization of the body. Depending on how and where a baby is born, they are colonized by different types of microbes. The first dose of microbes comes from the mother. Babies delivered vaginally are covered in a film of microbes as they pass through the birth canal (Aagaard et al, 2012). Included in the mix are bacteria that help babies digest their first meal. Babies delivered by caesarean section are colonized mainly by skin microbes—a very different set of species. Babies pick up microbes not only from their mothers, but also from every person and thing they touch. So, babies who are born at home are exposed to different types of microbes than babies born in hospitals (Torrazza and Neu, 2011). These very first differences—vaginal vs. c. section birth, home vs. hospital—are still measurable months and possibly even years after birth. They may have even longer-lasting impacts on health (Azad et al, 2013).

Within a few years the body becomes covered in thousands of different species of microbescolonizing every millimetre of the organismboth inside and out. By the time a person begins to socialize in community groups such as school, he or she has acquired vastly different populations living in the different parts of the body. Even as adults and into old age, the microbiota continues to be a dynamic system.

A changing microbiome

Very young babies are covered in a uniform mixture of microbes representing relatively few species. Within weeks, as babies pick up more microbes from their family members and surroundings, the types of microbes living in different parts of the body start to specialize.

As species compete for space and resources, the ones that are best equipped to live in a particular environment survive. The dry, light, oxygen-rich environment of the skin, for example, harbours a different population of microbes than the dark, watery environment of the mouth (Grice et al, 2009). Even more subtle differences can shape microbial populations: different microbes live on the cheek than in the crease next to the nose, and there are differences even between the left and right hands.

The food babies eat also feeds their microbes, and different microbes grow best on different types of food. Babies who are breastfed, for example, tend to have different gut microbes than babies who are given formula. Beginning solid food also causes the population to shift.

The number of species continues to increase (from about 100 species in the gut of young infants to about 1,000 in adults), and its make-up continues to change (Grice et al, 2012). Some of these changes mirror changing nutritional needs. For example, young babies get the vitamin folate from gut microbes while adults get most of their folate from food. Babies have more folate-producing microbes, and adults have more folate-harvesting microbes.

Parents and family members still have a big influence on microbiome composition. While microbial populations vary a lot, even among family members, relatives and people living together tend to have more similar microbiota than strangers. Over time, the early influence of the mother is diluted: young children's microbiomes come to resemble their fathers' as much as their mothers'.

Microbial populations shift and change a lot during early childhood. There is also a lot of variation among individuals. Variation is highest during childhood, and it gradually decreases with age. Events like a fever, a course of antibiotics, or new foods can cause sudden shifts in the microbiome, with effects that may last for years or even a lifetime (Jakobsson et al., 2010).

By age three, a child's microbiome looks a lot like an adult's, and it becomes much more stable. It continues to change in response to events like illness, disease, antibiotic treatment, fever, stress, injury, and changes in diet. But the population still tends to shift back to a "baseline" state (Koenig et al 2011).

Major life events—like puberty, pregnancy, and menopause—can cause larger shifts. For example, puberty affects skin microbes by causing changes in skin oils. Pregnancy brings on changes in the vaginal microbiome: species start to grow that will colonize and benefit the baby as it is born.

Senior citizens have a distinct microbial profile, with certain species becoming more common and others becoming less common. After age 65, the number of microbial species decreases, and the microbial populations become more similar among individuals (Yatsunenko et al., 2012).

Imbalance of intestinal microbial flora

Most people understand that the food that makes up their diet determines the health of their gut. In very general terms, eating junk food leads to declining gut health, while eating more vegetables leads to better gut health. Additionally, avoiding sugar is especially helpful in shifting the gut bacteria mix toward more health-promoting bacteria. However, what many people do not realize is that gut health can be influenced by more than just food. Some of the largest factors that influence gut health include the medicine, vitamins, and supplements that people take each day as well as the modern lifestyles that tend to impose stress on the various body systems.

The modern diet and its effect on the microbiota

The consumption of food containing live and beneficial microorganisms has dramatically reduced, and consequently, the developing mucosal immune systems are faced with different microflora that previous human populations.

While many people are making a conscious effort to eat for better gut health, there are several common mistakes made in the daily diet that may be compromising these well-intended efforts. Here are examples of such mistakes:

- Eating low-quality yogurt: Most yogurts have very few live bacteria in them and are packed with sugar and are made from poor-quality, low-fat sources of milk. Thus, the yogurt feeds the sugar-dependent bacteria in the gut, and the quantities of live probiotics are so small as to be inconsequential.
- Drinking too much kombucha: People drink too much kombucha or fermented drinks that can sometimes worsen candida, which wreaks havoc on gut health.
- Consuming too many "gut-healthy" processed foods: Processed food is still processed food. While these packaged products promise "probiotic health," they are often packed with disguised harmful ingredients such as processed yogurts with tons of sugar, probiotic drinks, and even sugar-heavy kombucha.
- Eating when stressed: When there are high levels of stress, the body diverts blood flow and energy away from the stomach, which slows down digestion. Food can then ferment in the stomach, leading to gas, bloating, and an imbalance of gut bacteria. Elevated cortisol also contributes to hormone imbalance, insulin resistance, suppresses thyroid function (can slow bowel movements and increase buildup of toxins), and create "leaky gut" by loosening the gut's tight junctions. Ideally, cooking and smelling food promotes healthy digestion as it starts in the mouth with the creation of amylase, one of

the digestive enzymes, while chewing food allows the brain to receive signals of satiation.

- Consuming too many raw vegetables: For many people with gut health issues resulting in diminished enzyme production and an altered gut microbiome, digesting raw vegetables can be quite difficult, leading to gas, bloating, and abdominal pain. This can be alleviated by eating cooked vegetables, and over time incorporate more raw vegetables as digestion improves. Another route, which is great for rebuilding the gut microbiome is to eat fermented veggies (like sauerkraut, kimchi) as part of a balanced diet.

Increases in observed incidence and severity of allergies and conditions such as IBS in the Western world have been linked with increases in standards of hygiene and sanitation, which have occurred concomitantly with decreases in thc number and range of infectious challenges encountered by the growing and developing host. This lack of immune education impairs the development of the immune system and allows the host to over-react to non-pathogenic antigen-containing commensal flora, resulting in inflammatory damage, allergy and/or autoimmunity (O'Sullivan et al., 2005). To combat these trends directly, the World Health Organization currently advocates the implementation of alternative disease control strategies, such as exploiting the prophylactic and therapeutic potential of probiotic bacteria (Dunne et al., 2009). Most of these probiotic microorganisms, isolated from such sources as feces of healthy individuals, are safe for human consumption and are available over the counter. Because of continued skepticism of such products, European Union funded research groups including medical, scientific and industrial interests, have agreed on criteria for selection and assessment of probiotics.

Health and the microbiome

Several health conditions are more frequently being associated to an imbalance of the microbiota. Although this is a topic highly related to functional foods, the discussion of specific diseases and the corresponding microbial agents goes beyond the scope of this book. Hence, a brief highlight is provided as a reference for the reader and to highlight the high frequency of microbial imbalances at a later section of this book.

Leaky gut syndrome

Leaky gut syndrome is not generally recognized by conventional physicians, but evidence is accumulating that it is a real condition that affects the lining of the intestines. The theory is that leaky gut syndrome (also called increased intestinal permeability), is the result of damage to the intestinal lining, making it less able to protect the internal environment as well as to filter needed nutrients and other biological substances. Consequently, some bacteria and their toxins, incompletely digested proteins and fats, and waste not normally absorbed may "leak" out of the intestines into the blood stream. This triggers an autoimmune reaction, which can lead to gastrointestinal problems such as abdominal bloating, excessive gas and cramps, fatigue, food sensitivities, joint pain, skin rashes, and autoimmunity.

The causes of leaky gut syndrome may include:

- chronic inflammation
- food sensitivity
- damage from taking large amounts of nonsteroidal anti-inflammatory drugs (NSAIDS)
- cytotoxic drugs
- radiation
- certain antibiotics
- excessive alcohol consumption
- compromised immunity.

Leaky gut syndrome may trigger or worsen such disorders as Crohn's disease, celiac disease, rheumatoid arthritis, and asthma.

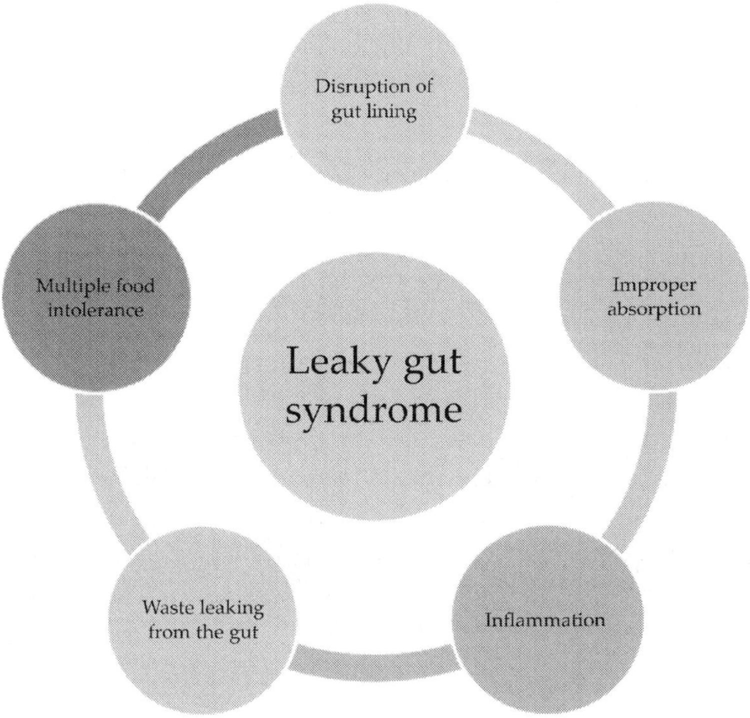

Figure 4: Diagram of leaky gut syndrome.

However, the main culprits associated with a leaky gut are food, infections, and toxins. Gluten and other inflammatory foods like dairy or toxic foods, such sugar and excessive alcohol, are suspected. The most common infectious causes are candida overgrowth, intestinal parasites, and small intestine bacterial overgrowth (SIBO). Toxins come in the form of medications, like Motrin, Advil, steroids, antibiotics, and acid-reducing drugs, and environmental toxins like mercury, pesticides and BPA from plastics. Stress and age also contribute to a leaky gut.

Some of the signs and symptoms of a leaky gut include:

- Digestive issues such as gas, bloating, diarrhea or irritable bowel syndrome (IBS).
- Seasonal allergies or asthma.
- Hormonal imbalances

- Diagnosis of an autoimmune disease such as rheumatoid arthritis, Hashimoto's thyroiditis, lupus, psoriasis, or celiac disease.
- Diagnosis of chronic fatigue or fibromyalgia.
- Mood and mind issues such as depression, anxiety
- Skin issues such as acne, rosacea, or eczema.
- Diagnosis of candida overgrowth
- Food allergies or food intolerances

Functional foods and leaky gut syndrome

Functional foods may help manage leaky gut. Once the body has patched up the leaks in the gut, help is needed to grow a healthy layer of favourable bacteria that help protect the GI tract and assist with digestion. These beneficial bacteria strengthen the immune system, improve metabolism, help the body make vitamins, and aid in the absorption of minerals. Probiotics may be highly beneficial for the management of leaky gut syndrome. These type of functional food is discussed in detail at a later chapter of this book.

In addition to probiotics, another functional food group to include in the management of leaky gut syndrome are fermented foods. Fermented foods will help repopulate and maintain the microbiota. These include foods such as kefir, yogurt (dairy or nondairy), sauerkraut, tempeh, and kombucha. Other perks of fermented foods include lowered inflammation, increased blood-sugar control, and improved antioxidant status. The only way to make a robust, permanent impact on gut flora, short of a fecal transplant, is with dietary changes.

Table 5: Common health conditions and their suggested microbiotic connection

Health condition	Description	Microbiotic connection
Acne	Acne may not be particularly dangerous, but it is certainly both unpleasant and very common. Acne typically happens with P. acnes, a type of bacteria that also lives on healthy skin, invades hair follicles, causing irritation and infection. In severe cases, acne is treated with antibiotics. Some of these drugs are applied directly to the skin, and others are taken as a pill. Delivered in pill form, antibiotics travel throughout the entire body, where they can also kill beneficial microbes. Some doctors are beginning to wonder if treating acne this way makes us vulnerable to other illnesses.	At the root of acne seems to be an imbalance between acne-causing bacteria and microbes that protect us. A better acne treatment may one day come in the form of a cream that nurtures the healthy microbes that protect our follicles from infection.
Antibiotic-associated diarrhea	People who take antibiotic pills to clear up an infection are sometimes left with the unfortunate after-effects of antibiotic-associated diarrhea. Antibiotics kill not only the bacteria that make us sick, but also the bacteria that keep us healthy. Antibiotics can seriously disturb the balance of our microbial populations.	To treat antibiotic-associated diarrhea, many have turned to probiotic foods like yogurt and bacteria-containing pills. But the types of bacteria in these foods and supplements tend to make up just a tiny proportion of the microbes in our intestines. Despite advertising touting their health benefits, supermarket probiotics haven't been very well studied at all. One treatment that has been shown to be effective against extremely severe imbalances in the intestinal ecosystem is the fecal transplant. Yes, doctors take poop from a healthy family member and deliver it to the sick person's intestine—very often leading to complete recovery within days. To take away some of the "yuck" factor, a few companies are developing techniques for growing complex mixtures of intestinal bacteria in the lab.

Allergies	Allergies happen when the immune system over-reacts to something that is normally harmless. Allergies are rare in developing countries, but in the United States, rates of asthma and hay fever have tripled over the last 30 years. Our genes haven't changed, so the increase must have something to do with the environment, quite possibly including our microbes.	Could hand sanitizer, air filters, and anti-microbial cleaners be the problem? Some evidence suggests that when kids are exposed to bacteria, viruses, and parasites—even some that make us sick—the immune system learns how to tell the difference between things that can hurt us and things that can't. Kids living in very clean environments may miss out on important immune system training. Additional training happens in the gut, where microbes interact closely with the immune system. Shifts in the food we eat and increases in antibiotic use affect gut microbes, possibly in a way that is harmful to the immune system.
Atherosclerosis	Hardening of the arteries, also called atherosclerosis, happens when a layer of sticky plaque builds up on the inside walls of blood vessels. A thick layer of plaque leaves less room for blood to flow, and tissue can become starved for oxygen. When pieces of plaque break off, they can become stuck in smaller blood vessels, completely blocking blood flow. People with hardening of the arteries often have no symptoms until something serious happens, like a heart attack or stroke	Decades ago, doctors noticed that many of their heart-attack patients also had red, irritated gums. Later, researchers found that plaque in blood vessels contains types of bacteria that normally live in the mouth. The best guess is that bacteria are entering the blood stream through blood vessels in unhealthy gums near infected teeth. The connection between plaque on teeth and plaque in blood vessels raises some interesting questions. Does good dental care prevent hardening of the arteries? Could scanning dental plaque for certain types of microbes diagnose disease before patients are struck down by heart attack or stroke? Could treating dental plaque also clear away blood-vessel plaque? There are many areas to explore.

Autism	The search for the cause of autism has turned up several weak genetic links and many suspected environmental triggers. Now researchers have learned that microbes may also be involved.	A few small studies have shown that children with autism have different microbes living in their intestines than children without the disorder. This difference could be explained by the often-picky eating habits of autistic children. Or the difference could be a side effect of an underlying genetic or environmental trigger—autistic children commonly have gastrointestinal problems. Still, the connection is intriguing. The causes of autism are varied and complex, and the same measures may prevent the disease in some but not in others. By learning about the connection between autism and microbes, we have the potential to find more ways to fight the disorder.
Autoimmune diseases	In people with autoimmune diseases, the immune system becomes confused and starts attacking the body's own tissues. With multiple sclerosis, the immune system attacks the protective wrappings around nerves and brain tissue. With rheumatoid arthritis, it attacks connective tissue in the joints. Dozens of other autoimmune diseases involve other organs and tissues throughout the body.	As with asthma and allergies, autoimmune diseases may be more likely to happen when the immune system doesn't get the proper training during childhood. Much of this training happens in and around the large intestine, and microbes are a big part of the process. Certain types of microbes living in the gut may protect us from autoimmune diseases, and others seem to make us more vulnerable. Mice that have segmented filamentous bacteria living in their gut are more affected by arthritis and encephalomyelitis, an autoimmune disease of the brain and spinal cord. Also in mice, certain microbes protect against type-1 diabetes, in which immune cells attack the insulin-producing cells of the pancreas.

Cancer	A disease caused by an uncontrolled division of abnormal cells in a part of the body	There is some evidence that microbes may have an indirect role in cancer. For example, some microbes cause inflammation, which appears to increase cancer risk. Several microbes also have more direct connections to cancer. H. pylori bacteria, a common cause of stomach ulcers, is also associated with cancers of the stomach and esophagus. A number of viruses cause cancer as well: Hepatitis B and C greatly increase the risk of liver cancer; the sexually transmitted HPV (human papilloma virus) is associated with many cervical cancers; and EBV (Epstein-Barr virus, the cause of mononucleosis) is associated with lymphomas (cancerous white blood cells) and other cancers.
Cavities	You probably know that eating lots of sticky, sugary foods leads to cavities. But did you know that microbes were involved? Dental plaque, the sticky layer that forms on our teeth when we don't brush, contains hundreds of species of microbes. A few of these are bacteria that use sugar as food and release acid as a byproduct. The acid eats through the hard enamel on the outside of our teeth, causing cavities. Bacteria can then grow inside our teeth, leading to infection. When we brush our teeth, we scrub away the microbes.	People with lots of cavities tend to have certain species of mouth bacteria, and people with healthy teeth tend to have others. It turns out that even though we have lots of different kinds of bacteria in our mouths, just a few cause cavities. In the future, we may be able to buy a mouthwash that targets these bacteria, protecting us from tooth decay. Meanwhile, the good news is that we know how to protect ourselves from getting cavities: cut down on sugar, and brush teeth at least twice every day.

Depression	Depression is a mood disorder that causes a persistent feeling of sadness and loss of interest. Also called major depressive disorder or clinical depression, it affects how you feel, think and behave and can lead to a variety of emotional and physical problems.	We are only beginning to understand the connections between our microbes and behavior. Doctors have long known that certain infectious microbes can influence behavior: the rabies virus causes insomnia, agitation, and fear of water; the protist Toxoplasma gondi takes away rodents' fear of cats. So, it shouldn't have been surprising when researchers found that some microbe-related intestinal disorders are associated with anxiety, depression, and other mental symptoms. Interestingly, skin appears to be part of this connection as well. Patients with the skin condition acne rosacea have a higher incidence of depression than the general population, and some evidence suggests that swallowing certain types of bacteria may improve both the acne and the depression.
Diabetes	The hormone insulin signals our cells to take up sugar (glucose) from our blood. People with type 2 diabetes are less responsive than normal to insulin—they are "insulin resistant." Without treatment, blood sugar levels can become dangerously high, gradually damaging the kidneys, eyes, blood vessels, and nerves.	Type 2 diabetes is much more common today than it was few decades ago, and its rise may have come from our shift to a high-fat, high-sugar diet. Changing what we put into our guts can shift the population of microbes that live there.

Eczema	Eczema is an allergic skin condition that leaves sufferers with patches of red, itchy, scaly skin. Like with other allergies, eczema may result from an improperly trained immune system.	Patches of skin with eczema have different microbes living on them than healthy skin. Researchers don't know if these differences are causing eczema or if the eczema is causing differences. But even if the microbes aren't the cause, they may offer opportunities for treatment. We know that certain microbes produce compounds that keep our skin healthy. Future eczema treatments may come in the form of a cream that nurtures beneficial microbes, or maybe even a cream that contains some of the microbes themselves.
Gastric ulcers	Gastric ulcers are painful sores inside the stomach or the first part of the small intestine. A healthy digestive tract has a protective layer of mucus, but ulcers are open to attack from the stomach acids that help digest our food. In the early 1980s, doctors discovered that the vast majority of gastric ulcers (around 80%) are associated with Helicobacter pylori bacteria. To treat severe ulcers, patients are often given mixtures of two to three different types of antibiotics.	H. pylori lives in about half of all healthy people as well, and most of these people don't develop ulcers. We don't understand why some people are more susceptible, but it looks like diet plays a role. Eating different types of food changes the abiotic factors of the stomach, making conditions more or less hospitable to different types of bacteria. Interestingly, H. pylori may also be important for developing a healthy immune system. People who have antibodies against H. pylori, showing that they have been exposed to this bug at some point in their lives, are less likely than others to develop asthma. And those that do develop asthma tend to do so at a later age.

Inflammatory bowel disease	Inflammatory bowel diseases (including Crohn's disease, Ulcerative colitis, and others) are a group of disorders that affect the intestines. These often-painful conditions leave sufferers with vomiting, diarrhea, and anemia (low iron). In the worst cases, surgeons must remove damaged sections of intestine.	Interactions between microbes and the cells that line the intestine appear to be important in inflammatory bowel disease. Some types of bacteria release molecules that irritate the intestine, loosen the connections between cells, and allow large molecules to leak through the intestinal wall into the blood stream. Some people with Crohn's disease also have a genetic variation that changes how their cells interact with microbes. The variation is in the NOD2 gene, which codes for a protein that recognizes and helps to kill certain types of bacteria when they get uncomfortably close to the cells that line the intestine. A defect in this mechanism allows bacteria to invade the intestinal wall, where they can cause inflammation and infection. Future treatments may focus on discouraging the growth of bacteria that cause inflammation and encouraging the growth of ones that suppress it.
Malnutrition	A lack of proper nutrition, caused by not having enough to eat, not eating enough of the right things, or being unable to use the food that one does eat.	Malnutrition isn't simply a matter of lacking calories and nutrients. Some people eat enough nutrients but can't absorb them. One clue about the role of microbes came from looking at identical twins, one twin undernourished and the other one not. The twins had the same genes, and they ate the same food—but they had different gut microbiota. Our microbial communities are established during our first few years, and they influence our health for life. If we learn more about the relationship between microbes and nutrition, we may be able to help babies grow healthy microbes right from the start. By manipulating the microbiome, doctors may be able to help patients take in more nutrients from essentially the same food.

Obesity	An excess of body weight due to an excess of body fat.	Thin people and obese people tend to have different types of microbes living in their guts. And when obese people lose weight, their microbiota change accordingly. With rates of obesity and related health problems increasing at an alarming rate, it has become an area of intense research. Obesity is highly heritable, and it probably involves genes that pass from parent to child. And body weight is clearly influenced by diet, so the environment and personal habits also play a role. However, several studies have also turned up evidence linking obesity to the microbiome: A diet high in fat, sugar, and simple carbs is bad for the "healthy" gut microbes that keep us thin while it encourages the growth of "unhealthy" microbes that make us obese. Obese individuals harbor microbes that are better at extracting energy from food, as well as microbes that signal the body to store energy as fat. Bacteria transplanted from overweight mice to thin mice make the thin mice gain weight. The relationship among genetics, the environment, and the microbiome as it relates to obesity is certainly complex. But while the genome is fixed and habits are hard to change, the microbiome is changeable. Could a nutritional supplement help healthy microbes grow?

Food selection and the microbiome

Diet affects the microbiome by increasing or decreasing the population of some bacteria relative to others and by introducing new bacteria into the microbiome (Rosenberg and Zilber-Rosenberg, 2016). A person's dietary history begins with how they are fed as new-borns. Human milk is not only the best source of nutrition for a new-born; it is also a source of microbes and prebiotic oligosaccharides, indigestible by a baby's gastrointestinal tract enzymes, that serve as carbon and energy sources for the beneficial bacteria in the baby's gut (Barile and Rastall, 2003). Once the breastfeeding process ends, and weaning begins, the introduction of solid foods (usually beginning at about 6 months of age)results in large changes in the microbiome. The gut microbiota of children no longer breast-fed becomes enriched in bacteria that are prevalent in adults, such as Lachnospiraceae, Roseburia, Clostrium, and Anaerostipes, rather than Bifidobacterium and Lactobacillus which dominated the gut microbiota of breast-fed infants (Backhed et al., 2015).

The human gut microbiome can rapidly respond to changes in the diet This is particularly useful if one considers the diversity of human dietary lifestyles. Not only long-term dietary intake influences the composition of the gut microbiome, but also short-term changes in the diet may alter the microbial community structure in a rapid and predictable manner. These changes can occur over a holiday period due to celebratory foods or during a trip taken abroad with the exposure to foreign microbes. In a clinical study, David et al. (2014) were able to induce, in just five days, differences in the microbiota, that would be metabolically more fit to the type of diet administered, entirely animal or entirely plant products. An animal-based diet increases the abundance of bile-tolerant microorganisms, such as Alistipes, Bilophila and Bacteroides, and decreases the levels of Firmicutes that metabolize dietary plant polysaccharides, such as Roseburia, Eubacterium rectale and Ruminococcus bromii. Furthermore, volunteers placed on a three-day high or low-calorie diet, showed that in the case of the energy increase, even this short-term was associated with an increased Firmicutes/Bacterioidetes ratio (Jumpertz et al., 2011). This point is of particular interest for people following a cycling-type of diet such as intermittent fasting or one of the many existing detox programs that alter the normal eating habits for a few days at a time.

Diet can not only bring about variation in the microbiome by altering the relative abundance of pre-existing microbiota, but can also change it by introducing novel bacteria. Foods that are not cooked contain numerous and diverse live microorganisms that can become a source of novel bacteria. Some of them may become established in the gut and provide health benefits. In 1907, the Nobel laurate Metchnikof hypothesized that ingestion of lactic-acid bacteria in the form of yogurt enhances longevity (1907). Evidence is now available that yogurt consumption is beneficial for intestinal health by restoring normal gut microbiota and suppressing inflammation (Pei et al., 2015). In addition to yogurt, foods that contain large numbers of beneficial bacteria include miso soup, sauerkraut, kombucha tea, dark chocolate, pickles, kimchi and kefir.

Credit: Raul Garduño Diaz 2017

Figure 5: Foods to boost the human microbiome.

Considering the microbiome as a distinct biological entity and a unit of selection in evolution, interaction between the normal microbiota and the host generally leads to improved wellness. Although the idea that resident microbial communities are important contributors to the health and wellness of their hosts is not new. Scientists are now developing a broader and deeper appreciation of the wide range of interactions occurring between the host and the microbiome. In humans, symbiotic microbes protect against pathogens (Huppert et al., 1953; Guarino et al., 2009; Witkin et al., 2013), provide nutrients (including vitamins and amino acids (Sekirov et al., 2010)), prime the adaptive or antigen-specific immune system during development and into adulthood (Lee and Mazmanian, 2010; Clarke et al., 2010), regulate postnatal angiogenesis (Stappenbeck et al., 2002), play a key role in energy homeostasis and obesity (Backhead et al., 2004; Turnbaugh et al. 2006; Vrieze et al., 2012), affect human behaviour (Bercik et al., 2011; Diaz et al., 2011; Foster and McVey 2013; Kang et al., 2013) and sometimes may contribute to inflammation (Kostic et al., 2014), and diseases such as diabetes (Hartsra et al., 2015).

In most animals, including humans, the gut harbours the largest bulk of the microbiome, which naturally interacts in a complex manner with the diet. Each component of the diet can, in principle, cause a change in the microbiome and different microbiomes can have different effects on a certain diet. The changes resulting from these interactions may lead to specific and sometimes novel physiological effects.

WHAT ARE FUNCTIONAL FOODS?

Definition

Functional foods are those that provide a health benefit beyond that of energy and nutrients such as vitamins and minerals. The term may also apply to organoleptic traits purposely bred into existing products. An example of the former is the naturally high polyphenol content found in coffee, which will be discusses in a later chapter; while the latter may refer to the enhanced edible colours of farmed crops, such as purple potatoes or golden rice, having been enriched with anthocyanin and carotenoid contents respectively (Marion Bliss, 2014). Functional foods may be "designed to have physiological benefits and/or reduce the risk of chronic disease beyond basic nutritional functions, and may be similar in appearance to conventional food and consumed as part of a regular diet" (USDA, 2010).

Historical background

The term "functional foods" was first used in Japan in the 1980s. If the most restrictive definition of functional foods is employed, the functional food movement began in Japan during the 1980's, when the Japanese government launched three major research initiatives designed to identify health-enhancing foods to control the rising cost of medical care. In 1991, a regulatory framework, Foods for Special Health Uses, was implemented, identifying those ingredients expected to have specific health benefits when added to common foods, or identifying foods from which allergens had been removed. FOSHU products were to be in the form of ordinary food (not pills or sachets) and consumed regularly as part of the diet. Initially, eleven categories of ingredients were identified for which sufficient scientific evidence indicated beneficial health effects. The Japanese Ministry of Health recognized foods containing these ingredients as functional foods. They were intended to improve intestinal function, reduce blood lipids and blood pressure, enhance calcium or iron absorption or serve as non-cariogenic sweeteners. In addition, low phosphorus milk was approved for people with renal insufficiency and protein-modified rice for people with rice globulin allergy. This pioneer approval processwas the first type of regulation put into place in regards to functional foods.

Figure 6: Seal for FOSHU approval.

FOSHU refers to foods containing ingredient with functions for health and officially approved to claim its physiological effects on the human body. FOSHU is intended to be consumed for the maintenance / promotion of health or special health uses by people who wish to control health conditions, including blood pressure or blood cholesterol. In order to sell a food as FOSHU, the assessment for the safety of the food and effectiveness of the functions for health is required, and the claim must be approved by the Ministry of Health, Law and Welfare of Japan.To date, two claims for the reduction of disease risk have been approved and received FOSHU approval. These are:

1) Calcium and osteoporosis: "Intake of proper amount of calcium contained in healthy meals with appropriate exercise may support healthy bones of young women and reduce the risk of osteoporosis when aged."

2) Folic acid and neural tube defects: "Intake of proper amount of folic acid contained in healthy meals may support women to bear healthy baby by reducing the risk of neural tube defect, such as spondyloschisis, during foetal development (MHLW, 2017).

If the broadest, least restrictive definition is employed, the use of functional foods for promoting health and relieving symptoms is as old as the practice of what is now called traditional medicine. Specific dietary recommendations for treating or preventing various types of illness have been documented in Hippocratic and Vedic texts and the canons of traditional Chinese medicine. Traditional Chinese remedies frequently contain recipes for combining specific foods with culinary and non-culinary herbs to produce healing mixtures. Folk medicines, from the East and West, have always depended upon functional foods. Very simple examples of this include: peppermint (Mentha piperita) tea has a long history of use for digestive complaints. Peppermint oil contains spasmolytic components that block calcium channels in smooth muscle. Cranberry (Vaccinium macrocarpon) juice contains proanthocyanidins that inhibit the attachment of E.coli to the epithelium of the urinary bladder, explaining its efficacy in prevention of bacterial cystitis and its traditional use for treatment of urinary infection.

Another way to illustrate the historical existence of functional foods is through herbs. Herbs and spices are added to food to enhance flavour and in addition were initially used to inhibit spoilage and maintain the innocuity of various foods. Many of these herbs and spices have documented medicinal uses that render them functional foods when broadly defined. For example, thyme (Lamiaceae spp) was used as a vermifuge in ancient Egypt. Today it is well known that thyme oils possess potent antimicrobial properties. Ginger (Zingiber officinale root), cinnamon (Cinnamomum spp bark) and licorice (Glycyrrhiza glabra root) are common ingredients in Chinese herbal tonics and have been widely used in Western folk medicine for treating digestive disorders. Ginger contains over four hundred biologically active constituents. Some have antimicrobial, anti-inflammatory, or anti-platelet effects; others enhance intestinal motility, protect the intestinal mucosa against ulceration and dilate or constrict blood vessels. Cinnamon oil contains cinnamaldehyde and various phenols and terpenes with antifungal, anti-diarrheal, vasoactive and analgesic effects. Recent research has identified phenolic polymers in cinnamon with insulin-sensitizing and insulin-mimetic activity, leading to the recognition that regular consumption of cinnamon may help prevent type 2 diabetes. The most studied component of licorice, glycyrrhizin, inhibits the enzyme beta-hydroxysteroid dehydrogenase type 2, potentiating the biological activity of endogenous cortisol. Glycyrrhizin also inhibits the growth of Helicobacter pylori. Glycyrrhizin and its derivatives may account for the anti-inflammatory and anti-ulcerogenic effects of licorice.

Fermentation is a form of food modification initially developed for the preservation of food and drinks. The health-enhancing effects of fermented foods have a place in folk medicine and are now being "rediscovered." Several fermented foods have health benefits that exceed those of their parent foods and hence can be considered functional foods when the concept is broadly defined. These include red wine, yogurt and tempeh.

Red wine is a whole fruit alcohol extract that concentrates the naturally occurring polyphenols found primarily in the seed and skin of different types of grapes. Its consumption is associated with protection against heart disease, perhaps because red wine polyphenols inhibit the production of free radicals and lipid peroxides that result from the simultaneous ingestion of cooked meat.

Fresh yogurt, when it contains live cultures of lactic acid-producing bacteria, can prevent the development of traveller'sdiarrhoea, antibiotic-induced diarrhoea, rotavirus infection, and vaginal yeast infection, decrease the incidence of post-operative wound infection following abdominal surgery and restore the integrity of the intestinal mucosa of patients who have received radiation therapy.

Tempeh is made from dehulled, cooked soybeans fermented by the fungus, Rhizopus oligosporus. Not only is its protein content higher than the parent soybean, it has antibiotic activity in vitro and the ability to shorten childhood diarrhoea in vivo.

Modification of a food to make it less harmful by removing potential toxins or allergens may also create a functional food. Using this criterion, infant formula, protein hydrolysates, low sodium salt substitutes, low fat dairy products and low erucic acid rapeseed oil (canola oil) might all alsobe considered functional foods.

More recently, interest in the development of functional foods has quickly spread from Asia to North America and Europe, where the concept was expanded to include any food or food component providing health benefits in addition to its basic nutritive value.

In Europe, functional food proponents distinguished functional foods from dietetic foods, which are defined by law. European dietetic foods are intended to satisfy special nutritional requirements of specific groups rather than to enhance physiologic function or prevent disease through non-nutritive influences. They include infant formula, processed baby foods (weanling foods), low-calorie foods for weight reduction, high-calorie foods for weight gain, ergogenic foods for athletes, and foods for special medical purposes like the treatment of diabetes or hypertension.

In the United States, functional food proponents have distinguished functional foods from medical foods, defined by law as special foods designed to be used under medical supervision to meet nutritional requirements in specific medical conditions.

In both domains, functional foods have been viewed as whole foods or food components with the potential for preventing cancer, osteoporosis or cardiovascular disease; improving immunity, detoxification, physical performance, weight loss, cognitive function, and the ability to cope with stress; inhibiting inflammation, free-radical pathology and the ravages of aging; and modulating the effects of hormones. Researchers have sought to validate biomarkers that demonstrate functional improvement in response to dietary intervention, identify the chemical components of functional foods responsible for those effects and elucidate the mechanism of action of those components. The scientific validation of all proposed r possible health claims is a major objective for both scientists and the food industry.

In China, functional foods (referred to as health foods) have been viewed as part of an unbroken medical tradition that does not separate medicinal herbs from foods. Over 3000 varieties of health foods are available to Chinese consumers, most derived from compound herbal formulas for which the active ingredients and their mechanism of action are unknown, all claiming multiple effects on various body systems, with little experimental evidence for safety and efficacy but widespread acceptance due to their history of use.

MICROORGANISMS IN FUNCTIONAL FOODS

Probiotics

The wrong bacteria in the wrong place can cause problems, but the right bacteria in the right place can have benefits. Probiotics are live bacteria and yeasts that are good for health, especially the digestive system. A growing amount of scientific research suggests that it may be possible to treat and even prevent some illnesses with foods and supplements containing certain kinds of live bacteria. Northern Europeans consume a lot of these beneficial microorganisms, called probiotics (from pro and biota, meaning "for life"), because of their tradition of eating foods fermented with bacteria, such as high-quality yogurt. Probiotic-loaded beverages are also big business in Japan.

When it comes to the health benefits of probiotics, promoting a healthy digestive tract and a healthy immune system are their most widely studied benefits now. Although more research is needed, there is encouraging evidence that probiotics may help:

- Treat diarrhoea, especially following treatment with certain antibiotics
- Prevent and treat vaginal yeast infections and urinary tract infections
- Treat irritable bowel syndrome
- Speed treatment of certain intestinal infections
- Prevent or reduce the severity of colds and flu
- Maintain a stable weight
- Clear up skin conditions

There are several different kinds of probiotics, and their health benefits are determined by the job that they do in the gut. The most well-known ones to date include lactobacillus and bifidobacteria.

What are some food sources of probiotics?

The discovery of the benefits of probiotics is likely to have begun with sour milk. Today we have many other options to get various bacteria from our foods, although it is not as simple as just adding them to the food. For there to be health benefits, the microorganism must be able to survive the passage through the gastrointestinal tract, survive the food manufacturing process, and grow and survive during the ripening or storage period. In addition, the bacteria must not negatively affect product quality and be included on the Generally Recognized as Safe (GRAS) list.

That being said, some of the best food sources to obtain these bacteria include:

- Kimchi (fermented cabbage)
- Yogurt
- Miso (fermented soybean paste)
- Tempeh (fermented soybean in patty form)
- Sauerkraut (fermented cabbage)
- Aged soft cheese
- Sourdough bread
- Sour pickles

Considering that the microbiome as a combination of beneficial and harmful bacteria, it is important to have enough of the beneficial type to ensure a proper functioning of the human body. Scientists are learning more each day about the role of probiotics and other functional foods in keeping people healthy.

Probiotics as a group may contain a variety of living microorganisms. The most common are bacteria that belong to groups called Lactobacillus and Bifidobacterium. Each of these two broad groups includes many types of bacteria. Other bacteria may also be used as probiotics, and so may various types of yeasts such as Saccharomyces boulardii.Prebiotics are not the same as probiotics. The term "prebiotics" refers to dietary substances, mostly non-digestible carbohydrates, that favour the growth of beneficial bacteria over harmful ones. The term "symbiotic" refers to products that combine probiotics and prebiotics.

- Lactobacilli: Lactobacillus acidophilus, also known as L. acidophilus, is a beneficial bacterium highly desired for inhabitation of the small intestine. These microorganisms are very acid-resistant and tolerate bile quite effectively. Besides nestling into the mucosal lining of the small intestine, L. acidophilus can also be found colonizing and protecting the lining of the vagina and cervix in addition to the urethra. In these places, acidophilus will work to produce lactic and other organic acids inhibiting the adherence and multiplication of pathogenic microbes.

L. acidophilus antagonizes a collection of pathogens including Escherichia coli (E. coli, the cause of "Montezuma's Revenge"), Staphylococcus aureus and Salmonella (common causes of food poisoning), Candida albicans (common causes of yeast infections and syndromes), and other pathogens such as Shigella, Clostridium, Listeria, and Helicobacter species. Yogurt contains the probiotic lactobacillus acidophilus, but many people cannot digest conventional yogurt. Kefir may prove to be a better source of this beneficial bacteria.

- Bifidobacteria: The bifidos were originally identified in the faeces of a breast-fed infant. These good bacteria are some of the first microbes to take up residence in the gastro-intestinal (GI) tract of newborns. They also dominate the intestines in breast-fed babies. This is one reason why the status of probiotics in the mother in pregnancy is very important. As babies age and are weaned, the bifidobacteria numbers and species change as the diet changes. Typically, their numbers continue to decrease with age.

Bifidobacterium, along with Lactobacillus, are some of the most well-known probiotics and are an important part of the microbiota. These beneficial bacteria are lactic-acid producing bacteria (LAB), just like Lactobacillus, and they both produce vitamins, bacteriocins (antibacterial chemicals) and antibiotic-like substances. Both have significant health benefits on the digestive and immune systems. Unlike many Lactobacillus, however, bifidobacteria also produce acetic acid (a short-chain fatty acid). The significance of this is that acetic acid is more effective at reducing the growth of yeasts and moulds than is lactic acid. Acetic acid can also be used as energy by the human body. Producing both lactic and acetic acids and other beneficial compounds makes bifidobacteria the probiotic most qualified to be in the vagina and urogenital tract, where yeasts and opportunistic bacteria can cause infections. In addition, their presence in the colon is highly beneficial, where the opportunity for disease to emerge is greater due to slower faecal transit time. This probiotic species also produces ethanol and other acids, can produce vitamins, and can help with lactose intolerance.

In addition to the GI tract benefits, some members of this species have a beneficial effect on the organism just by being inside the body since they are able to cause immune system modifications. They can help increase immune function in some cases and have anti-inflammatory actions in other cases. Furthermore, the mere presence of some of these probiotics can cause other microbes that live in the intestines to perform helpful duties that they otherwise would not do. This suggests that bifido can communicate with other beneficial bacteria to help maintain a healthy status.Bifidobacterium is on the European QPS (Qualified Presumption of Safety) list and is generally regarded as safe (GRAS).

Data from the 2012 National Health Interview Survey (NHIS) reported that approximately 4 million (1.6 percent) U.S. adults had used either a probiotic or a prebiotic in the past 30 days. Amid adults, probiotics or prebiotics were reported to be the third most commonly used dietary supplement other than vitamins and minerals. Data from the same survey also showed that the use of probiotics quadrupled between 2007 and 2012. The 2012 NHIS also showed that 300,000 children age 4 to 17 (0.5 percent) had used probiotics or prebiotics in the 30 days prior to the survey.

Although some probiotics have shown promise in laboratory studies, strong scientific evidence to support specific uses of probiotics for most health conditions remainswanting. The U.S. Food and Drug Administration (FDA) has not approved any probiotics for the prevention of or treatment of any health problem. Some experts have cautioned that the rapid growth in marketing and use of probiotics may have outpaced scientific research for many of their proposed uses and benefits.

Probiotic safety

Whether probiotics are likely to be safe as a form of functional food depends on the initial and overall state of health of the consumer. In people who are generally healthy, probiotics have a good safety record. Side effects, if they occur at all, usually consist only of mild digestive symptoms such as gas.

On the other hand, there have been reports linking probiotics to severe side effects, such as dangerous infections, in people with serious underlying medical problems. The people who are most at risk of severe side effects include critically ill patients, those who have had surgery, very sick infants, and people with weakened immune systems. This last point must be strongly considered as severe side effects are not likely to trouble the general population.

Nevertheless, even for healthy people, there are uncertainties about the safety of probiotics. Because many research studies on probiotics have not looked closely at safety, there is not enough information at the moment to answer certain safety questions. Most of the knowledge about safety comes from studies of Lactobacillus and Bifidobacterium; less is known about other probiotics. Information on the long-term safety of probiotics is limited, and safety may differ from one type of probiotic to another. For example, even though a National Centre for Complementary and Integrative Health (NCCIH)-funded study showed that a particular kind of Lactobacillus appears safe in healthy adults age 65 and older, this does not mean that all probiotics would necessarily be safe for people in this age group.

BIOACTIVE INGREDIENTS IN FUNCTIONAL FOODS

Clinical trial data, as well as evidence from epidemiological in vivo and invitro studies, indicates that a plant-based diet can reduce the risk of chronic disease, particularly certain types of cancer and other chronic conditions. For example, a review of 200 epidemiological studies (Block et al., 1992) showed that cancer risk in people consuming diets high in fruits and vegetables was only one-half that in those consuming lessof these foods. It has been suggested that there are components in a plant-based diet, other than traditional nutrients, that can reduce cancer risk. Steinmetz and Potter (1991) identified more than a dozen classes of these biologically active plant chemicals, now known as "phytochemicals."Phytochemicals are naturally occurring bioactive components of plants. Some phytochemicals are thought to promote human health but have no known requirement and thus can be an important component of functional foods (USDA, 2010). These physiologically active compounds are produced via secondary metabolism in relatively small amounts. Until fairly recently, they have been generally assumed to be irrelevant, sometimes even harmful, to human health. The number of identified phytochemicals has increased dramatically in the last decade. Some groups of phytochemicals which have or appear to have significant health potentials are: carotenoids, phenolic compounds (flavonoids, phytoestrogens, phenolic acids), phytosterols and phytostanols, tocotrienols, organosulfur compounds (allium compounds and glucosinolates), and nondigestible carbohydrates (dietary fibre and prebiotics).

Antioxidants

The potential for supplements and functional foods, in terms of development of new products and the available growing markets, is considerable. Currently, there are accepted, clinically demonstrated, uses for some dietary supplements, for example:

- The use of folic acid in pregnant women to prevent possible neural tube defects in the foetus.
- The use of B-vitamin supplements to counteract the malabsorption from the diet caused by treatment with biguanide (Metformin) in diabetics.
- Inclusion of specific plant sterols within a healthy diet in order to reduce the overall cholesterol content, particularly levels of low-density lipoprotein (LDL)-cholesterol.
- The use of fish oils in the diet in order to treat and prevent heart disease and arthritis, as well as perhaps controlling metabolic syndrome (e.g. obesity, non-insulin dependent diabetes, abnormal blood lipid patterns and high blood pressure) and improving concentration in younger children.

From a consumer point of view, the idea that supplements offer protection from chronic illnesses, from the common cold to various types of cancer, has caught on. Functional foods appear to have been accepted under a similar 'body of scientific evidence'. However, results from human studies ten to be insufficient for their use as hard evidence, when compared to encouraging results obtained from in vitro studies.

Relatively few studies have examined the effects of antioxidants at levels likely to be relevant in vivo to nutritional studies on healthy people using validated biomarkers. By implication, the diverse group of compounds known as antioxidants may have the capacity to regulate multiple key cellular processes such as cell proliferation and apoptosis, with important implications for understanding major disorders such as coronary heart diseases and cancer, and their prevention. However, although oxidative compounds have been frequently reported to cause cellular damage, they may also be mediators of change (e.g. cell signals) in cell function, proliferation and differentiation. Thus, there is still a need to understand the relative importance of the disparate roles of oxidants as well as the action of so-called antioxidants on oxidant behaviour and cellular response. To do this, however, requires a change of approach. It is no longer simply a matter of how dietary antioxidants affect the cellular process but also how cellular processes interact with food components at the molecular level. Nutritional genomics – how food compounds interact with our DNA, RNA and proteins – has tremendous potential to change the future of dietary guidelines regarding recommendations for the individual.

Nutrigenomics

Nutrigenomics is the study of how foods affect genes and how individual genetic differences can affect the way we respond to nutrients (and other naturally occurring compounds) in the foods we eat. Nutrigenomics has received much attention recently because of its potential for preventing, mitigating, or treating chronic disease, and certain cancers, through small but highly informative dietary changes. The conceptual basis for this new branch of genomic research can best be summarized by the following five tenets of nutrigenomics:

- Under certain circumstances and in some individuals, diet can be a serious risk factor for a number of diseases.
- Common dietary chemicals can act on the human genome, either directly or indirectly, to alter gene expression or structure.
- The degree to which diet influences the balance between healthy and disease states may depend on an individual's genetic makeup.
- Some diet-regulated genes (and their normal, common variants) are likely to play a role in the onset, incidence, progression, and/or severity of chronic diseases.
- Dietary intervention based on knowledge of nutritional requirement, nutritional status, and genotype (i.e., "personalized nutrition") can be used to prevent, mitigate or cure chronic disease.

The promise of nutritional genomics is personalized medicine and health based upon an understanding of individual nutritional needs, nutritional and health status, and genotype. Even with the use of post-genomics technologies, there are still enormous challenges to be faced in understanding how food or food compounds affect health. The technologies are relatively new, and researchers are rethinking their approaches to procedures such as study design, sample collection, data storage, and volunteer selection for practical, theoretical and, not least, logistical reasons. One factor is the volume of data that these techniques generate. For example, a single analytical procedure using, for example DNA microarrays, can easily generate tens of thousands of data points, each relating to the expression of a single gene, some of which will be linked – often in unknown ways – whilst others are independent. These data are also derived from only a single method of analysis and results should be confirmed, albeit on a more limited scale, with at least one alternative approach. Whilst clinical symptoms of a disease are biomarkers in themselves, they are not suitable for early detection and therefore prevention. A series of biomarkers, in tissues that are easily accessible from apparently healthy individuals, as well as those at different stages of disease progression, is desirable. Early biomarkers can yield information on the nature of ROS damage and the action of food compounds – including antioxidants – in vivo. Initial results have indicated that genomic 'fingerprints' can be obtained, which can be used as biomarkers (Tang et al. 2001; Walker et al. 2002). However, more work is needed to determine whether these patterns can serve as valid biomarkers in relation to consumption of certain food compounds for the maintenance of good health or prevention of disease. The majority of work undertaken to date has typically been performed using surrogate tissues. The validity of using such tissues has yet to be demonstrated, as does control of the effects of processing techniques (e.g. post-genomic technologies) on these tissues; such effects could include changes in gene expression (e.g. inducing heat-shock proteins). Another issue facing nutrition research in the post-genomic era is the prospect of working from the point of ignorance. Although the human genome has now been mapped and considerable headway has been made in interpreting the information contained therein, the implications of subtle differences in the DNA code between individuals is not, as yet, understood. We have also come to understand that some apparently 'abnormal' genotypes may confirm some other benefit to the individual, e.g. protection from malaria. When armed only with the DNA code or with very limited understanding of the interaction between a given genotype and its effect on human health, it is difficult to make accurate predictions regarding risk of disease or benefit from

Carotenoids

Carotenoids are a class of more than 750 naturally occurring pigments synthesized by plants, algae, and photosynthetic bacteria (Wang, 2014). These richly coloured molecules are the sources of the yellow, orange, and red colours of many plants. Fruit and vegetables provide most of the 40 to 50 carotenoids found in the human diet. α-Carotene, β-carotene, β-cryptoxanthin, lutein, zeaxanthin, and lycopene are the most common dietary carotenoids (Wang, 2014). α-Carotene, β-carotene and β-cryptoxanthin are provitamin A carotenoids, meaning they can be converted by the body to retinol. Lutein, zeaxanthin, and lycopene are non-provitamin A carotenoids because they cannot be converted to retinol.

For dietary carotenoids to be absorbed intestinally, they must be released from the food matrix and incorporated into mixed micelles (mixtures of bile salts and several types of lipids). Food processing and cooking help release carotenoids embedded in their food matrix and increase intestinal absorption (Wang, 2014). Moreover, carotenoid absorption requires the presence of fat in a meal. As little as 3 to 5 g of fat in a meal appears sufficient to ensure carotenoid absorption (Jalal, 1998 and van Het Hof, 2000), although the minimum amount of dietary fat required may be different for each carotenoid. The type of fat (e.g., medium-chain vs. long-chain triglycerides), the presence of soluble fibre, and the type and amount of carotenoids (e.g., esterified vs. non-esterified) in the food also appear to influence the rate and extent of carotenoid absorption (Pryadarshani, 2015). Because they do not need to be released from the plant matrix, carotenoid supplements (in oil) are more efficiently absorbed than carotenoids in food (van Het Hof, 2000).

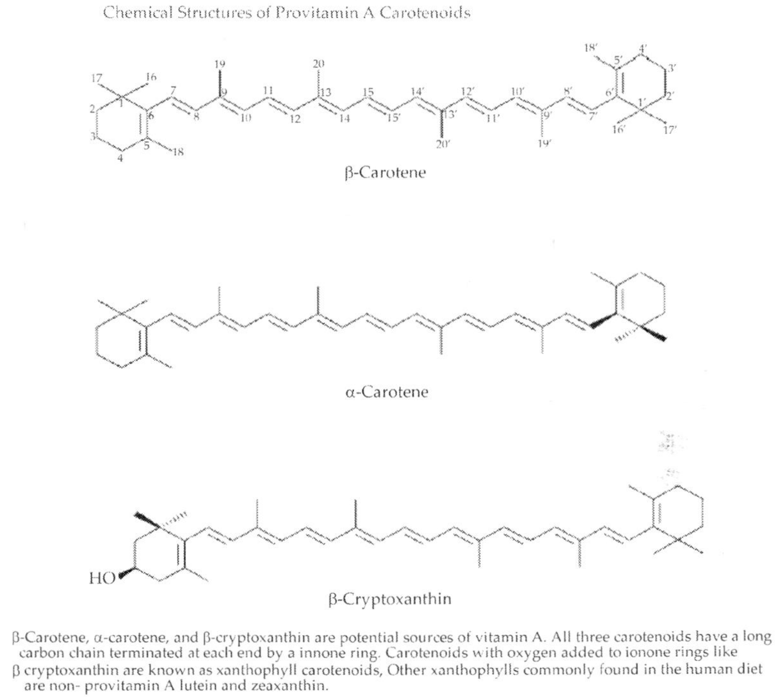

Figure 7: Chemical structures of provitamin A carotenoids.

Dietary carotenoids are thought to provide health benefits in decreasing the risk of disease, particularly certain cancers and eye disease. The carotenoids that have been most studied in this regard are beta-carotene, lycopene, lutein, and zeaxanthin. In part, the beneficial effects of carotenoids are thought to be due to their role as antioxidants. Beta-carotene may have added benefits due its ability to be converted to vitamin A. Furthermore, lutein and zeaxanthin may be protective in eye disease because they absorb damaging blue light that enters the eye. Food sources of these compounds include a variety of fruits and vegetables, although the primary sources of lycopene are tomato and tomato products. Additionally, egg yolk is a highly bioavailable source of lutein and zeaxanthin. These carotenoids are available in supplement form. However, intervention trials with large doses of beta-carotene found an adverse effect on the incidence of lung cancer in smokers and workers exposed to asbestos. Until the efficacy and safety of taking supplements containing these nutrients can be determined, current dietary recommendations of diets high in fruits and vegetables are advised.

Food sources of carotenoids

The most prevalent carotenoids in the human diet are α-carotene, β-carotene, β-cryptoxanthin, lycopene, lutein, and zeaxanthin (FNB, 2000). Most carotenoids in foods are found in the all-trans form, although cooking may result in the formation of other isomers. The relatively low bioavailability of carotenoids from most foods compared to supplements is partly due to the fact that they are associated with proteins in the plant matrix (Yeum and Russell, 2002). Chopping, homogenizing, and cooking disrupt the plant matrix, increasing the bioavailability of carotenoids (van Het Hof, 2015). For example, the bioavailability of lycopene from tomatoes is substantially improved by heating tomatoes in oil (Gartner et al, 1997 and Stahl and Sies, 19992). α-Carotene and β-carotene are provitamin A carotenoids, meaning they can be converted in the body to vitamin A. The vitamin A activity of β-carotene in food is 1/12 that of retinol (preformed vitamin A). Thus, it would take 12 μg of β-carotene from food to provide the equivalent of 1 μg (0.001 mg) of retinol.

Table 6:α-carotene content of selected foods

Food	Service	α-Carotene (mg)
Pumpkin, canned	1 cup	11.7
Carrot juice, canned	1 cup (8 fl oz)	10.2
Carrots, cooked	1 cup	5.9
Carrots, raw	1 medium	2.1
Mixed vegetables, frozen, cooked	1 cup	1.8
Winter squash, baked	1 cup	1.4
Plantains, raw	1 medium	0.8
Collards, frozen, cooked	1 cup	0.2
Tomatoes, raw	1 medium	0.1
Tangerines, raw	1 medium	0.09
Peas, edible-podded, frozen, cooked	1 cup	0.09

Table 7:β-carotene content of selected foods

Food (mg)	Service	α-Carotene
Carrot juice, canned	1 cup (8fl oz)	22.0
Pumpkin, canned	1 cup	17.0
Spinach, frozen, cooked	1 cup	13.8
Sweet Potato, baked	1 medium	13.1
Carrots, cooked	1 cup	13.0
Collards, frozen, cooked	1 cup	11.6
Kale, frozen, cooked	1 cup	11.5
Turnip greens, frozen, cooked	1 cup	10.6
Pumpkin pie	1 piece	7.4
Winter squash, cooked	1 cup	5.7
Carrots, raw	1 medium	5.1
Dandelion greens, cooked	1 cup	4.1
Cantaloupe, raw	1 cup	3.2

Adapted from USDA, 2015

The vitamin A activity of α-carotene from foods is 1⁄24 that of retinol, so it would take 24 μg of α-carotene from food to provide the equivalent of 1 μg of retinol. Orange and yellow vegetables like carrots and winter squash are rich sources of α- and β-carotene. Spinach is also a rich source of β-carotene, although the chlorophyll in spinach leaves hides the yellow-orange pigment. Some foods that are good sources of α-carotene and β-carotene are listed in Table 6 and Table 7 (USDA, 2015).

Lutein

In the past, lutein – a carotenoid or pigment found in many fruits and vegetables – was primarily known as a nutrient for good eye health. But new research shows that it may also play a strong role in developing and maintaining brain function, and can enhance learning and memory.

For infants, researchers are learning that lutein in breastmilk supports the development of brain function. In fact, research suggests that as babies get more lutein, either from formula or from breastmilk (when their mother has a high-lutein diet), there is more lutein detected in their brains. Increased levels of lutein may help speed up cognitive milestones in infants. These cognitive milestones could include an infant making eye contact with a parent, moving their head to look at something in their peripheral vision, or turning their attention to someone when they hear their name.

School-age children may see higher test scores when they eat more lutein. Emerging research suggests the amount of lutein present in preadolescent children's eyes directly relates to their brain development and learning skills, which can predict performance on tests.

College students also can reap the benefits of this super nutrient. Baseline data and additional follow-up research suggest that lutein and zeaxanthin—another carotenoid very similar to lutein—supplementation improves processing speed and efficiency, along with memory in young and older adults. For older adults, lutein supports mental sharpness and healthy aging.

Zeaxanthin

Zeaxanthin is a carotenoid commonly found in nature. Zeaxanthin is primarily synthesized by plants, although some micro-organisms also produce it. The name "zeaxanthin" is a combination of the Greek root words "zea" and "xanthos." Zea is the word for grain and is also the taxonomic genus that includes corn. Zanthos is the Greek word for yellow, referring to zeaxanthin's colour.

The best dietary sources of zeaxanthin are generally leafy, green vegetables such as collard greens, kale, spinach, romaine lettuce, Swiss chard and turnip greens. Additional common vegetables that contain a high level of zeaxanthin include goji berries, broccoli, corn, kiwifruit, peas and zucchini. Zeaxanthin is also found in other foods like eggs and spirulina.

Zeaxanthin is an essential component of the macula, where it helps to filter out blue light from the sun. This light has a high-energy level, which can damage the eyes. Humans can't biosynthesize zeaxanthin, so they must obtain it through dietary sources and supplements. The lenses in the eye must remain clear to collect light and focus it on the retina. Oxidation of the lens often occurs over time, which can make the lenses cloudy. Zeaxanthin has antioxidant properties that may help keep the lenses clear in the presence of free radicals.

Lycopene

Lycopene is a naturally occurring chemical that gives fruits and vegetables a red colour. It is one of a number of pigments called carotenoids. Lycopene is found in watermelons, pink grapefruits, apricots, and pink guavas. It is found in particularly high amounts in tomatoes and tomato products. In North America, 85% of dietary lycopene comes from tomato products such as tomato juice or paste. One cup (240 mL) of tomato juice provides about 23 mg of lycopene. Processing raw tomatoes using heat (in the making of tomato juice, tomato paste or ketchup, for example) changes the lycopene in the raw product into a form that is easier for the body to use. The lycopene in supplements is about as easy for the body to use as lycopene found in food.

People take lycopene for preventing heart disease, "hardening of the arteries" (atherosclerosis); and cancer of the prostate, breast, lung, bladder, ovaries, colon, and pancreas. Lycopene is also used for treating human papilloma virus (HPV) infection, which is a major cause of uterine cancer. Some people also use lycopene for cataracts and asthma.

Tocotrienols

Tocotrienols are unsaturated analogs of tocopherol (vitamin E). They occur in the unsaponifiable fraction of vegetable oils; rice bran oil and palm oil are the most significant sources (Eitenmiller, 1997). A number of plant foods, ranging from kale and broccoli, to cereal grains and nuts have also been found to contain tocotrienols (Piironen et al., 1986). There are at least four known forms of tocotrienol, with γ-tocotrienol as the main and most potent cholesterol-lowering form (Parker et al., 1993). The cholesterol-lowering effect of tocotrienols is attributed to their ability to inhibit hydroxymethylglutaryl-CoA (HMG-CoA) reductase, the rate-limiting enzyme in the cholesterol synthesis pathway (Parker et al., 1993). It has been suggested that α-tocopherol attenuates the inhibitory effect of γ-tocotrienol (Qureshi et al., 1996). Tocotrienols have also been demonstrated to possess vitamin E activity (Tan, 1989), antioxidant activity (Serbinova et al., 1993) and antitumor properties (Komiyama et al., 1992). It was hypothesized that the inhibition of HMG-CoA reductase by tocotrienols also results in suppression of tumour growth (Elson and Qureshi, 1995).

Tocotrienols may offer many important complementary benefits to the tocopherol form of vitamin E, such as:

1. Improving lipid profiles: Tocotrienols help lower plasma cholesterol levels by blocking HMG-CoA reductase. This enzyme is a rate-limiting step in producing cholesterol, so blocking it is an effective means of lowering cholesterol levels. In fact, the common statin drugs used for lowering lipid levels also operate by inhibiting HMG-CoA reductase, although they do so by a different mechanism. Human research shows that tocotrienol supplementation can lower total serum lipids by up to 23%, total cholesterol by 30%, and LDL (bad) cholesterol by 42%. Not all human studies show this dramatic reduction in LDL and cholesterol.22 And animal studies demonstrate significant reductions in lipids, with the added benefit of reducing lipid oxidation; a precursor to the arterial damage that leads to stroke and heart attacks. Tocotrienols also can lower dangerous forms of apolipoproteins, which are lipid-carrying proteins that can increase cardiovascular risk.
2. Reducing Homocysteine: Tocotrienols have also been found to reduce blood levels of homocysteine, another important contributor to cardiovascular disease. When scientists used a tocotrienol-rich fraction in a rat study to reduce plasma homocysteine and heart oxidative stress, they discovered that it proved more powerful than folate, the standard homocysteine-reducing vitamin. In rabbits, tocotrienol supplementation after a high-fat diet significantly lowered a host of markers of both inflammation and heart muscle damage.
3. Protecting against ischemia: If a vessel does become blocked, tocotrienols offer potent resistance to the loss of blood flow known as ischemia, which is generally involved in the development of heart attack or stroke. This was powerfully demonstrated in a study of rabbits, in which supplementation not only lowered their lipids, but also protected their heart muscle from damage by ischemia induced in the laboratory.

This was especially true of the alpha- and gamma-tocotrienol, which reduced the area of heart muscle damage significantly.

4. Guarding against stroke-related brain damage: Acute ischemic stroke (reduced blood flow to an area of the brain) remains a leading cause of death and disability worldwide. Oral tocotrienols have been shown to protect against and minimize stroke-related brain damage in animal models. This protection is the result of several independent mechanisms.
5. Improving Metabolic Syndrome: In an exciting look, ahead at what the future may hold in human studies, preclinical research has shown that tocotrienols can reverse many changes seen in metabolic syndrome, including improving lipid profiles, reducing atherosclerotic lesions, lowering blood glucose and markers of glycation, normalizing blood pressure, and inhibiting new fat deposits. At the functional level, tocotrienols improve heart muscle function, improve glucose and insulin tolerance, and reduce infiltration of inflammatory cells into heart muscle

Organosulfur Compounds

Allium compounds are organosulfur compounds (OSCs) found in allium vegetables such as garlic, onion, scallion, chive, shallot and leek, which account for the distinctive flavour and aroma as well as the many reported medicinal effects of these vegetables. The OSCs in allium vegetables have been reported to exert various physiological effects, including (Huang et al., 1994):

- antimicrobial activity
- lipid-lowering effect
- hypocholesteremic activity
- antithrombic effect
- inhibition of platelet aggregation
- hypoglycemic activity
- lipoxygenase and tumour inhibition.

The glucosinolates are sulphur-containing glucosides prevalent in the cruciferous family of vegetables, especially the Brassicas (e.g. cabbage, broccoli, Brussels sprouts, cauliflower), and are also present at relatively high levels in oilseeds such as rapeseed and in condiments such as mustard seed.

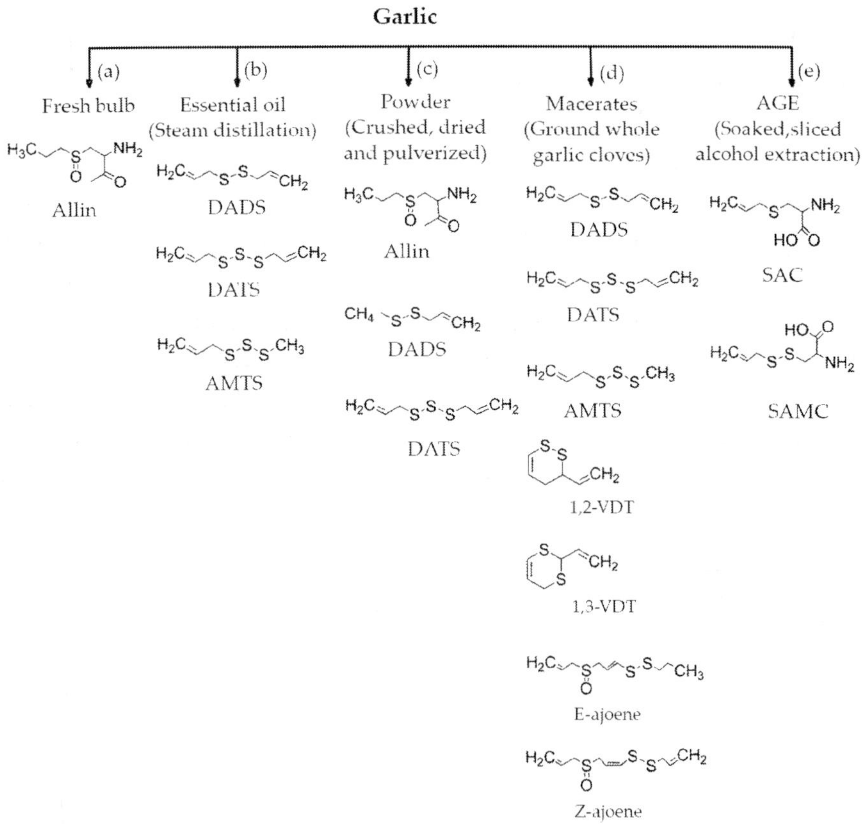

Figure 8: Organosulfur compounds in garlic.

Over a hundred different glucosinolates have been identified in the plant kingdom, but only about 10 are present in cruciferous vegetables (Stoewsand, 1995). Although the glucosinolates are structurally diverse, there are only three principal groups, based on the side-chain structure: aliphatic, aromatic and indolyl (heteroaromatic) glucosinolates (Mithen et al., 2000). All the glucosinolates have a β-D-thioglucose group, a sulphonated oxime moiety and a variable side-chain derived from methionine, tryptophan, phenylalanine or some branched-chain amino acids (Fenwick et al., 1983). When the plant tissue is damaged by food preparation or chewing, the glucosinolates are brought into contact with and are hydrolyzed by the endogenous enzyme, myrosinase, to yield a complex mixture of products mainly isothiocyanates, nitriles and thiocyanates. Glucosinolate breakdown products exert a variety of antinutritional and toxic effects in higher animals, the most thoroughly studied of which is the goitrogenic effect of some products (Heaney & Fenwick, 1995). At present, there is little or no epidemiological evidence of this goitrogenic effect causing a disease in humans.

Dietary Fibre

The generally accepted definition of dietary fibre is that of Trowell, that dietary fibre consists of the remnants of edible plant cells polysaccharides, lignin, and associated substances resistant to (hydrolysis) digestion by the alimentary enzymes of humans (Prosky, 2000). Dietary fibre can be considered a functional food when it imparts a special function to that food aside from the normal expected function and similarly when the dietary fibre is used as an additive to foods. For example, dietary fibre contributes to:

- colonic health
- bifidobacterial or lactobacillus stimulation in the gut
- coronary artery health
- cholesterol reduction
- glucose metabolism
- insulin response
- blood lipids
- cancer (Prosky, 2000).

Newer fibre materials (both soluble and insoluble) provide benefits beyond the traditional fibre concepts of basic digestive health. These include: cardiovascular health and immune system support, antioxidant benefits, regulation of normal blood sugar and blood lipid levels, sports and fitness nutrition, weight management, satiety and bone health.

Soluble fibre

Soluble fibre helps lower blood cholesterol levels and control blood sugar levels. This type of fibre is obtained from oats, barley, psyllium, oranges, dried beans and lentils. The health benefits of soluble fibre include:

- Heart protection: Inside the digestive system, soluble fibre attaches to cholesterol particles and takes them out of the body, helping to reduce overall cholesterol levels and the risk of heart disease. Oatmeal may offer the most heart protection.
- Diabetes protection: Because soluble fibre isn't well absorbed, it doesn't contribute to the blood sugar spikes that can put you at risk for type 2 diabetes and heart disease. If diabetes is already present (either type 1 or type 2) soluble fibre can even help keep the condition under control.
- Weight loss: Soluble fibre can also help you get to -- or stay at -- a healthy weight by maintain satiation without adding many calories to the diet.
- Healthy bowel movements: Soluble fibre soaks up water as it passes through the digestive system, which helps bulk up your stool and guard against constipation and diarrhoea. In fact, most fibre supplements contain mostly soluble fibre.

Insoluble fibre

Insoluble fibre helps promote regularity and a healthy digestive system. This type of fibre is obtained from wheat bran, whole grains, and some vegetables. The health benefits of insoluble fibre include:

- Weight loss: Like soluble fibre, insoluble fibre can play a key role in controlling weight by staving off hunger pangs.
- Digestive health: Eating lots of insoluble fibre also helps keeps bowel movements regular, and if constipation occurs, adding more of it to your diet can get things moving. Insoluble fibre can also improve bowel-related health problems, like constipation, haemorrhoids, and faecal incontinence.

Dietary fibres in functional foods

Companies in the functional food market utilize a range of 'nutritional' fibres extracted from plant food sources as dietary fibre supplements, such as Indian gooseberry fruits, cocoa beans, coconut endosperm, fenugreek, flax seeds and ginger rhizomes. Guar gum, acacia gum and pectin, which are derived from seed isolates, plant exudates and processed fruit by-products, are excellent sources of soluble dietary fibre. Yet another category comprises the synthetic cellulose, ethers such as sodium carboxyethyl cellulose and hydroxypropyl methyl cellulose. Both categories of soluble fibre are water soluble and serve a variety of functions such as viscosity enhancement, emulsion stabilization and freeze/thaw protection for new food products (Crawford, 2010).

Cereals, as a source of dietary fibre, offer an alternative for the production of functional foods. The multiple beneficial effects of cereals can be exploited in different ways leading to the design of novel cereal foods or cereal ingredients that can target specific populations. Cereals can be used as fermentable substrates for the growth of probiotic microorganisms. The main parameters that have to be considered are:

- the composition and processing of the cereal grain
- the substrate formulation
- the growth capability and productivity of the starter culture
- the stability of the probiotic strain during storage
- the organoleptic properties
- the nutritional value of the final product.

Additionally, cereals can be used as sources of nondigestible carbohydrates that besides promoting several beneficial physiological effects can also selectively stimulate the growth of lactobacilli and bifidobacteria present in the colon and act as prebiotics. Cereals contain water-soluble fibre, such as β-glucan and arabinoxylan, oilgosaccharides, such as galacto- and fructo-oligosaccharides and resistant starch, which have been suggested to fulfil the prebiotic concept. Separation of specific fractions of fibre from different cereal varieties or cereal by-products, according to the knowledge of fibre distribution in cereal grains, could be achieved through processing technologies, such as milling, sieving, and debranning or pearling. Finally, cereal constituents, such as starch, can be used as encapsulation materials for probiotics in order to improve their stability during storage and enhance their viability during their passage through the adverse conditions of the gastrointestinal tract. It could be concluded that functional foods based on cereals is a challenging perspective, however, the development of new technologies of cereal processing that enhance their health potential and the acceptability of the food product are of primary importance.

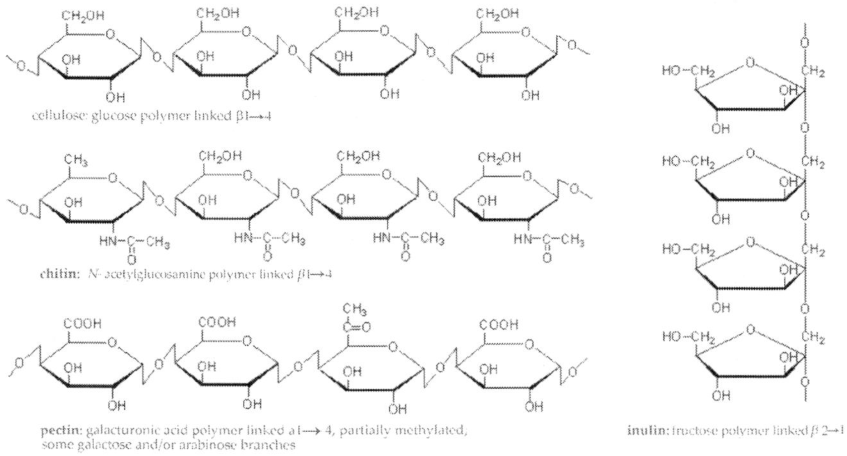

Figure 9: Chemical structure of various dietary fibres.

Inulin

Inulin, a relative newcomer to the fibre category, is a low molecular weight oligosaccharide of fructose. Inulin is a soluble fibre that serves as an excellent source of carbohydrates for probiotic organisms populating the lower portion of the gastrointestinal system. Dairy products with active cultures such as yogurt combined with inulin, and foods promoting inulin content, are growing as more consumers become aware of the benefits of probiotic maintenance.Inulin and oligofructose belong to a class of carbohydrates known as fructans. The main sources of inulin and oligofructose that are used in the food industry are chicory and Jerusalem artichoke. Inulin and oligofructose are considered as functional food ingredients since they affect physiological and biochemical processes in rats and human beings, resulting in better health and reduction in the risk of many diseases. Experimental studies have shown their use as bifidogenic agents, stimulating the immune system of the body, decreasing the levels of pathogenic bacteria in the intestine, relieving constipation, decreasing the risk of osteoporosis by increasing mineral absorption, especially of calcium, reducing the risk of atherosclerosis by lowering the synthesis of triglycerides and fatty acids in the liver and decreasing their level in serum. These fructans modulate the hormonal level of insulin and glucagon, thereby regulating carbohydrate and lipid metabolism by lowering the blood glucose levels; they are also effective in lowering the blood urea and uric acid levels, thereby maintaining the nitrogen balance. Inulin and oligofructose also reduce the incidence of colon cancer. The biochemical basis of these beneficial effects of inulin and oligofructose have been discussed. Oligofructoses are non-cariogenic as they are not used byStreptococcus mutans to form acids and insoluble glucans that are the main culprits in dental caries. Because of the large number of health promoting functions of inulin and oligofructose, these have wide applications in various types of foods like confectionery, fruit preparations, milk desserts, yogurt and fresh cheese, baked goods, chocolate, ice cream and sauces. Inulin can also be used for the preparation of fructose syrups.

Chicory

The plant that is most commonly used industrially for the extraction of inulin-type fructans belongs to the Compositae family, i.e., chicory. Native chicory inulin is a no fractionated inulin extracted from fresh roots. Because of the β-configuration of the anomeric C2 in its fructose monomers, inulin-type fructans resist hydrolysis by human small intestinal digestive enzymes, which are specific for α-glycosidic bonds. They have thus been classified as "nondigestible" oligosaccharides.

Chicory is a woody, herbaceous plant that has a wealth of health benefits, including the ability to ease digestive problems, prevent heartburn, reduce arthritis pains, detoxify the liver and gallbladder, prevent bacterial infections, boost the immune system, and reduce the chance of heart disease. It is also a natural sedative, and can protect against kidney stones, and benefit attempts to lose weight. All in all, this small plant is a powerful addition to any diet.

Chicory grows well in many different conditions, and is commonly found on the roadsides in Europe, where it originated, but it has now naturalized into the Americas and can be found widely there. The leaves, buds, and roots all have certain beneficial qualities, and the leaves are used in a similar way as spinach in many cultures. The root is often ground into a powder and used as a coffee substitute, which doesn't negate all of its health benefits. It has been used for hundreds of years as a herbal remedy for many diseases, and continues to prove its worth today.

In terms of nutritive value, chicory is a great source of vitamins and minerals, including zinc, magnesium, manganese, calcium, iron folic acid, and potassium, as well as vitamin A, vitamin B6, vitamin C, vitamin E, and vitamin K. These are the main agents behind all of the health benefits that come from chicory.

Organic chicory root contains inulin, as mentioned above. Many plants contain inulin, but chicory root has the highest concentration. Nourishing the healthy flora in the intestines enhances digestion. Furthermore, since inulin content is not digestible, the lack of glucose can help promote optimal blood sugar levels while also increasing stool bulk and consistency to help eliminate toxins efficiently. Chicory may also provide direct functional support to the digestive reactions in the body. Chicory root increases the flow of bile, which supports digestion. Because extra bile helps break down fats, chicory root may help optimize blood composition and is worth consideration by anyone seeking to achieve optimal liver and gallbladder health. These properties of the chicory plant and root make it a potential functional ingredient for the development of new products.

Jerusalem artichoke

Jerusalem artichokes are native to the Central America. The plant is technically an evergreen perennial but cultivated as an annual crop. Once established, it grows vigorously with multiple branches, reaching about 5-10 feet height, slightly taller than sunflower plant, and carries many golden-yellow flower heads at the terminal end of branches.

The plant bears many starchy edible rhizomes, firmly attached to stem underneath the soil surface. Jerusalem tubers feature grey, purple, or pink skin externally, and sweet, delicate textured ice-white flesh inside. Some roots have quite bumpy and extremely knobby outer surface, making their clean a tougher task. Each tuber weighs about 75 to 200 g.

It is one of the finest sources of dietary fibres, especially high in oligo-fructose inulin as mentioned above, which is a soluble non-starch polysaccharide. The soluble as well as insoluble fibres in this tuber add up to the bulk of food by retaining moisture in the gut. The tuber contains small amounts of anti-oxidant vitamins such as vitamin-C, vitamin-A, vitamin-E. These vitamins, together with flavonoid compound like carotenes, helps scavenge harmful free radicals and thereby offers protection from cancers, inflammation and viral cough and cold. Further, Jerusalem artichokes are an excellent source of minerals and electrolytes, especially potassium, iron, and copper. 100 g of fresh root holds 429 mg or 9% of daily required levels of potassium. Potassium is a heart friendly electrolyte which brings reduction in the blood pressure and heart rate by countering pressing effects of sodium. 100 g of fresh sunchoke contains 3.4 mg or 42.5% of iron, probably the highest amount of this trace element among the common edible roots and tubers. It also contains small levels of some of the valuable B-complex group of vitamins such as folates, pyridoxine, pantothenic acid, riboflavin, and thiamin.

Fatty Acids

A major part of the interaction between diet and microbiota in humans involves the conversion of plant-based dietary fibre and other non-digestible food ingredients into short chain fatty acids (SCFAs) by microbial fermentation in the colon. SCFAs, the end products of fermentation of dietary fibres by the anaerobic intestinal microbiota, have been shown to exert multiple beneficial effects on mammalian energy metabolism. The mechanisms underlying these effects are the subject of intensive research and encompass the complex interplay between diet, gut microbiota, and host energy metabolism. The principal SCFAs produced via bacterial fermentation are acetate, propionate, and butyrate, present in the human colon in the approximate molar ratio of 60:20:20 (Cummings et al., 1987). SCFAs show important and pleiotropic functional effects via their influence on key regulatory proteins. SCFA affect cell proliferation and function and microbe-to-host relationship, have anti-inflammatory, anti-tumorigenic properties and are major players in energy metabolism and maintenance and function of the immune system. Given the many effects of SCFAs, they also play a role in affecting various modern-day diseases including obesity, diabetes, inflammatory bowel diseases, and colorectal cancer (Tan et al. 2014).

SCFAs contribute up to 5-10% of the total energy in a healthy body (McNeil, 1984). Acetate, the main SCFA in the blood, plays a key metabolic role for peripheral tissues being a substrate for energy production, lipogenesis and cholesterol synthesis (Wolever et al., 1991). Recent research suggests that propionate reduces energy intake by attenuating reward-based eating behaviour in the striatum, the subcortical part of the forebrain that is a critical component of the reward system (Byrne et al., 2016). Butyrate production by gut microbiota acts as an important energy and carbon source for colonocytes, it is protective against colon cancer and acts also as an anti-inflammatory, antioxidative agent and plays a role in intestinal barrier functions (Leonel and Alvarez-Leite, 2012). Butyrate, via its effect on regulatory proteins which influences gene expression, has also multiple effects on brain function and behaviour (Stilling et al, 2016).

In the last few decades, it became apparent that SCFAs might play a key role in the prevention and treatment of the metabolic syndrome, bowel disorders, and certain types of cancer. In clinical studies SCFA administration positively influenced the treatment of ulcerative colitis, Crohn's disease, and antibiotic-associated diarrhoea. The molecular mechanisms by which SCFAs induce these effects are an active field of research.

The amount and type of fibre consumed has dramatic effects on the composition of the intestinal microbiota and consequently on the type and amount of SCFAs produced. For the host, the in vivo SCFA production rates as well as the intestinal SCFA concentrations on different fibres are most relevant. In humans, the effect of dietary fibre intake has been studied mainly by measuring the SCFA concentrations in faeces followed by calculating the total rate of SCFA excretion. Faecal secretion rates for SCFA are in the range of 10–30 mmol/day for diets with high fibre content compared with 5–15 mmol/day for control diets. In most studies, acetate is the predominant SCFA in the faeces, followed by propionate and butyrate. It is important to note that faecal SCFA concentrations do not reflect their concentration and production rate in the intestine as most SCFAs are taken up by the host and therefore faecal SCFA excretion provides little information about actual intestinal SCFA metabolism. For the production of SCFAs, it is important that the gut microbiota work as a community, but also that the gut microbiota have symbiotic associations with the host. The molecular H_2 that is produced during acetate formation must be used by other bacteria in the community to avoid accumulation of H_2 which would inhibit the ability of primary fermenters to oxidize NADH. The CO_2 that is needed in the primitive electron transfer chain is partly provided by the host. Humans produce on average about 0.7 kg of CO_2 per day. Part of that production is excreted into the lumen of the gut as HCO3− in exchange for SCFA anions. Most likely this is an important pH regulatory mechanism because protons in the lumen of the gut, formed during the production of SCFAs, are neutralized by bicarbonate under the formation of CO_2.

Because most SCFAs are absorbed by the host in exchange for bicarbonate, the luminal pH is the result of the microbial SCFA production and the neutralizing capacity of bicarbonate. As the concentration of SCFAs decline from the proximal to the distal colon, the pH increases from cecum to rectum. The drop in pH from the ileum to the cecum due to the higher SCFA concentrations has two effects. First, both in vitro and animal studies show that lower pH values change gut microbiota composition and second, it prevents overgrowth by pH-sensitive pathogenic bacteria like Enterobacteriaceae and Clostridia. Studies of human faecal microbial communities showed that at pH 5.5 the butyrate-producing bacteria such as Roseburia spp. and Faecal bacterium prausnitzii, both belonging to the Firmicutes phylum, comprised 20% of the total population. When fermentable dietary fibres become limiting in the more distal parts of the large intestine, the luminal pH increases to 6.5, the butyrate-producing bacteria almost completely disappear, and the acetate- and propionate-producing Bacteroides-related bacteria become dominant.

Interest has been recently rekindled in SCFAs with the emergence of functional foods aimed at improving colonic and systemic health. Dietary carbohydrates, specifically resistant starches and dietary fibre, are substrates for fermentation that produce SCFAs, primarily acetate, propionate, and butyrate, as end products. The rate and amount of SCFA production depends on the species and amounts of microflora present in the colon, the substrate source and gut transit time. SCFAs are readily absorbed. Butyrate is the major energy source for colonocytes. Propionate is largely taken up by the liver. Acetate enters the peripheral circulation to be metabolized by peripheral tissues. Specific SCFA may reduce the risk of developing gastrointestinal disorders, cancer, and cardiovascular disease. Acetate is the principal SCFA in the colon, and after absorption it has been shown to increase cholesterol synthesis. However, propionate, a gluconeogenerator, has been shown to inhibit cholesterol synthesis. Therefore, substrates that can decrease the acetate: propionate ratio may reduce serum lipids and possibly cardiovascular disease risk. Butyrate has been studied for its role in nourishing the colonic mucosa and in the prevention of cancer of the colon, by promoting cell differentiation, cell-cycle arrest and apoptosis of transformed colonocytes; inhibiting the enzyme histone deacetylase and decreasing the transformation of primary to secondary bile acids as a result of colonic acidification. Therefore, a greater increase in SCFA production and potentially a greater delivery of SCFA, specifically butyrate, to the distal colon may result in a protective effect. Butyrate irrigation (enema) has also been suggested in the treatment of colitis. More human studies are now needed, especially, given the diverse nature of carbohydrate substrates and the SCFA patterns resulting from their fermentation. Short-term and long-term human studies are particularly required on SCFAs in relation to markers of cancer risk. These studies will be key to the success of dietary recommendations to maximize colonic disease prevention.

Flavonoids

One of the main types of phytonutrients, abundant in almost all fruits and vegetables, are the flavonoid family. Along with carotenoids, they are responsible for the intense colours in fruits and vegetables. Flavonoids are the largest group of phytonutrients, with over 6,000 types of compounds. Some of the best-known flavonoids are quercetin and kaempferol.

Recently, the health benefits of diets rich in fruits and vegetables have been attributed to their high content of various flavonoids. Like other phytonutrients, flavonoids act mainly as strong antioxidants with benefits for the immune system and other anti-inflammatory pathways. Diets rich in flavonoid-containing foods are sometimes associated with cancer, neurodegenerative and cardiovascular disease prevention. However, it is not yet clear whether the flavonoids themselves are responsible or if there exists a synergetic effect with various other compounds. The latter answer being more likely due to the relatively limited understanding by scientists as to the complexity of food components.

Onions, tea, strawberries, kale, grapes, Brussels sprouts, citrus fruit, parsley, and many spices are just a few natural foods rich in flavonoids (Xlibris, 2014).

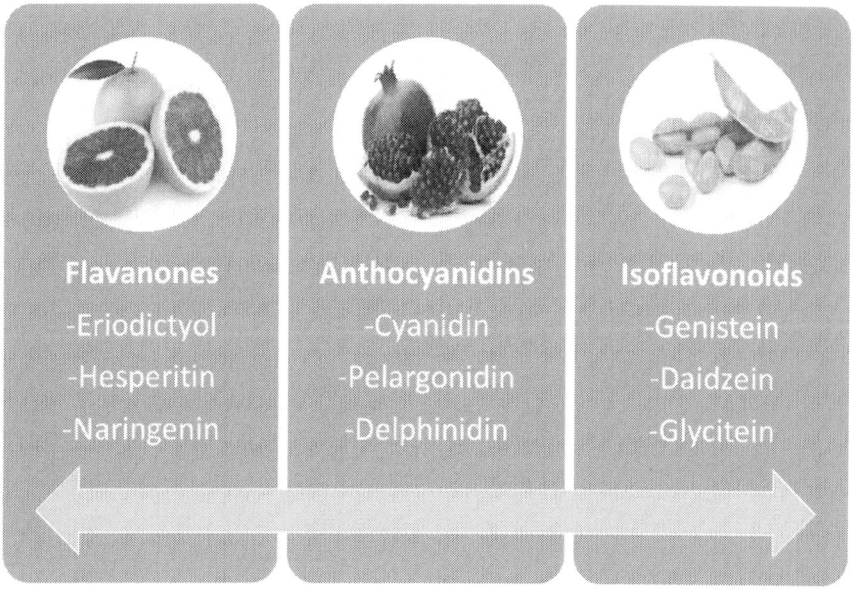

Figure 10: Flavonoids found in food.

There are several significant groups of flavonoids, including anthocyanidins, flavanols, flavones, flavonols, flavonones and isoflavones. Within the flavanol subgroup there are still more subgroups. Each of these subgroups and each type of flavonoid carries its own distinct set of actions, benefits and originating foods. Flavonoids are polyphenolic compounds that occur ubiquitously in foods of plant origin. Over 4000 different flavonoids have been described. They may have beneficial health effects because of their antioxidant properties and their inhibitory role in various stages of tumour development in animal studies. An estimation of the total flavonoid intake is difficult, because only limited data on food contents are available. It is estimated that humans ingest a few hundreds of milligrams per day. The intake of flavonols and flavones was inversely associated with subsequent coronary heart disease in most but not all prospective epidemiological studies. Flavonoids present in foods are considered non-absorbable because they are bound to sugars as β-glycosides. However, it has been found that human absorption of the quercetin glycosides from onions (52%) is far better than that of the pure aglycone (24%). Flavonol glycosides might contribute to the antioxidant defences of blood. Dietary flavonols and flavones content alone probably do not explain the cancer-protective effect of vegetables and fruits and a protective effect against cardiovascular disease is not conclusive. Below are the descriptions of the main groups, their sources and benefits:

- Flavones: Within the secondary metabolite class of flavonoids which consist of more than 9000 known structures, flavones define one of the largest subgroups. These include luteolin and apigenin. Their natural distribution is demonstrated for almost all plant tissues. The diverse functions of flavones in plants as well as their various roles in the interaction with other organisms offer many potential applications, not only in plant breeding but also in ecology, agriculture and human nutrition and pharmacology. In this context, the antioxidative activity of flavones, their use in cancer prevention and treatment as well as the prevention of coronary heart disease should be emphasized. The therapeutic potential of flavones makes these compounds valuable targets for drug design, including recombinant DNA approaches. The biosynthesis of flavones in plants was found to be catalysed by two completely different flavone synthase proteins (FNS), a unique feature within the flavonoids. The first, FNS I, a soluble dioxygenase, was only described for members of the Apiaceae family so far. The second, FNS II, a membrane bound cytochrome P450 enzyme, has been found in all other flavone accumulating tissues. This phenomenon is particularly of interest from the evolutionary point of view concerning the flavone biosynthesis and functions in plants. Recently, FNS I and FNS II genes have been cloned from a number of plant species. This now enables detailed biochemical and molecular characterizations and also the development of direct metabolic engineering strategies for modifications of flavone synthesis in plants to improve their nutritional and/or biopharmaceutical value. Good sources of flavones are celery, parsley, various herbs and hot peppers. Flavones are associated with overall antioxidant benefits and delaying the metabolizing of drugs.
- Anthocyanidins: Anthocyanins are the components in many fruits that are responsible for the blue, purple, and red colour of many plant tissues. These include malvidin, pelargondin, peoidin and cyanidin. The potential dietary intake of anthocyanins is fairly large (100+ mg/day) compared to other

plant phytochemicals due to their main and abundant food sources. The concentrations of anthocyanins in different fruits vary considerably as well as the individual anthocyanins present also are quite different in various fruits. Anthocyanins can provide protection against various forms of oxidative stress in animal models, however, most of the health-related responses have been observed at relatively high intakes of anthocyanins. Good sources of anthocyanidins include red, purple and blue berries; pomegranates; plums; red wine; and red and purple grapes. Anthocyanidins are associated with heart health, antioxidant effects and helping with obesity and diabetes prevention.

- Flavonones: These include hesperidin, eriodictyol and naringenin. Flavonones are found abundantly in citrus fruits. They are associated with cardiovascular health, relaxation and overall antioxidant and anti-inflammatory activity.
- Isoflavones: This subgroup includes genistein, glycitein and daidzein. Isoflavones are highly concentrated in soybeans and soy products, as well as legumes. They are phytoestrogens, meaning that they are chemicals that act like the hormone estrogen. Scientists suspect they may be beneficial in lowering the risk of hormonal cancers, such as breast, endometrial and prostate cancers, though study results are currently mixed. In various studies, isoflavones have sometimes acted as antioxidants and sometimes as oxidants, so their effect on cancer is unclear. They are also being studied as a way to treat menopausal symptoms.
- Flavonols: This widely distributed subgroup of flavonoids includes quercetin and kaempferol. They are found in onions, leeks, Brussels sprouts, kale, broccoli, tea, berries, beans and apples. Quercetin is an antihistamine associated with helping to relieve hay fever and hives. It is also known for its anti-inflammatory benefits. Kaempferol and other flavonols are associated with powerful anti-inflammatory and antioxidant activities leading to chronic disease prevention.
- Flavanols: There are three primary types of flavanols: monomers (more widely known as catechins), dimers and

polymers. Flavanols are found in teas, cocoa, grapes, apples, berries, fava beans and red wine. Catechins are especially common in green and white teas, while dimers, which are associated with lowering cholesterol, are found in black tea. Scientists suspect catechins might be useful in aiding chronic fatigue syndrome symptoms. Catechins are also associated with cardiovascular and neurological health. The beneficial effects of flavonoid consumption on cardiovascular risk are supported by mechanistic and epidemiologic evidence.

Many types of flavonoids are available in supplement form. While these may be a good option for those struggling to get sufficient fruits and vegetables in their diet, quercetin supplements and tea extracts may cause side effects such as nausea, vomiting, tremors and dizziness. There are no side effects of consuming flavonoids through plant-based foods. People who hope to consume flavonoids through plant-based foods should be aware that cooking and storing could change the flavonoid make up in fruits and vegetables. For example, onions stored at room temperature can lose up to one-third of their flavonoids in just two weeks. Up to 80 percent of some flavonoids can be lost in the cooking process. A good way to tell if your food is losing nutrients is by its colour; if its normally vivid colours start to fade while being boiled or cooked, your food is losing its phytonutrients.

Quercetin

Quercetin, the major representative of the flavonol subclass, is a strong antioxidant, and prevents oxidation of low density lipoproteins in vitro. It is ubiquitously present in foods including vegetables, fruit, tea and wine as well as countless food supplements and is claimed to exert beneficial health effects. This includes protection against various diseases such as osteoporosis, certain forms of cancer, pulmonary and cardiovascular diseases but also against aging. Especially the ability of quercetin to scavenge highly reactive species such as peroxynitrite and the hydroxyl radical is suggested to be involved in these possible beneficial health effects. Consequently, numerous studies have been performed to gather scientific evidence for these beneficial health claims as well as data regarding the exact mechanism of action and possible toxicological aspects of this flavonoid. Oxidized low density lipoproteins are atherogenic, and are considered to be a crucial intermediate in the formation of atherosclerotic plaques. This agrees with observations in epidemiological studies that the intake of flavonols and flavones was inversely associated with subsequent coronary heart disease.

The popularity of neosynthesized, naturally occurring plant extracts and phytochemicals to enhance mental and physical performance, maintain health, and prevent disease has increased dramatically in recent years due in part to the fact that they are accessed more easily, generally are less expensive, and are thought to have fewer side effects than manufactured drugs. The use of herbs and botanicals for these purposes has occurred for millennia, but not until recently have these substances been subjected to systematic experimental validation. Because the beneficial effects of a supplement like quercetin are largely dependent on its bioavailability after oral administration, the absorption, distribution, metabolism, and excretion of quercetin have been studied extensively in both laboratory animals and humans. Quercetin also has been shown to reach and accumulate in various tissues including colon, kidney, liver, lung, muscle, and brain, although the tissue distribution has not yet been studied in humans. Quercetin has GRAS according to criteria established by the U.S. Food and Drug Administration. There are no reports of harmful side effects in animals or humans at doses of several grams per day.

Kaempferol

Kaempferol is a polyphenol antioxidant found in fruits and vegetables. Many studies have described the beneficial effects of dietary kaempferol in reducing the risk of chronic diseases, especially cancer. Epidemiological studies have shown an inverse relationship between kaempferol intake and cancer. Kaempferol may help by increasing the body's antioxidant defence against free radicals, which promote the development of cancer. At the molecular level, kaempferol has been reported to modulate a number of key elements in cellular signal transduction pathways linked to apoptosis, angiogenesis, inflammation, and metastasis. Significantly, kaempferol inhibits cancer cell growth and angiogenesis and induces cancer cell apoptosis, but on the other hand, kaempferol appears to preserve normal cell viability, in some cases exerting a protective effect.

Kaempferol (3,5,7-trihydroxy-2-(4-hydroxyphenyl)-4H-1-benzopyran-4-one) is a flavonoid found in many edible plants (e.g. tea, broccoli, cabbage, kale, beans, endive, leek, tomato, strawberries and grapes) and in plants or botanical products commonly used in traditional medicine (e.g. Ginkgo biloba, Tilia spp, Equisetum spp, Moringa oleifera, Sophora japonica and propolis). Numerous preclinical studies have shown that kaempferol and some glycosides of kaempferol have a wide range of pharmacological activities, including antioxidant, anti-inflammatory, antimicrobial, anticancer, cardioprotective, neuroprotective, antidiabetic, anti-osteoporotic, estrogenic/antiestrogenic, anxiolytic, analgesic and antiallergic activities. The pharmacokinetics (e.g. oral bioavailability, metabolism, plasma levels) and safety of kaempferol are being studied. This information may help understand the health benefits of kaempferol-containing plants and may contribute to develop this flavonoid as a possible agent for the prevention and treatment of some diseases as a potential functional food ingredient.

Isothiocyanates

Isothiocyanates are sulphur-containing phytochemicals with the general formula R-NCS. Different molecules belong to this group. Isothiocyanates with the strongest anticancer effects are phenylethylisothiocyanate, benzylisothiocyanate and 3-phenylpropylisothiocyanate. Isothiocyanates occur naturally as glucosinolate conjugates in cruciferous vegetables. Isothiocyanates are also responsible for the typical flavour of these vegetables. Isothiocyanates can be found in cruciferous vegetables such as broccoli, cauliflower, kale, turnips, collards, Brussels sprouts, cabbage, radish, turnip and watercress. Glucosinolates are precursors of isothiocyanates. When the raw vegetables are chewed the plant, cells are broken and an enzyme (myrosinase) hydrolyses the glucosinolates into isothiocyanates. Since the early 1960's, both natural and synthetic isothiocyanates have attracted considerable and growing attention as important and effective protectors against chemical carcinogenesis in a number of animals. Some glucosinolates have been examined and appear to be very effective in inhibiting carcinogenesis. Evidence suggests that isothiocyanates exert their effects through a variety of distinct but interconnected signalling pathways important for inhibiting carcinogenesis, including those involved in detoxification, inflammation, apoptosis, and cell cycle and epigenetic regulation, among others. Tests performed in laboratories and in animal studies have shown that isothiocyanates:

- Inhibit the development of cancer by stopping cancer causing chemicals from becoming active.
- Play an important role in protecting our cells from damaging distortions.
- Have been found to control the over activity of cancer cells when added to them in a test tube.
- Have been found to induce appropriate cell death in cancer cells.
- Have been found to have an anti-inflammatory activity (important for protection against heart attacks).
- Have been found to inhibit the development of chemically induced cancers of the lung, liver, stomach, small intestine, colon, and breast.

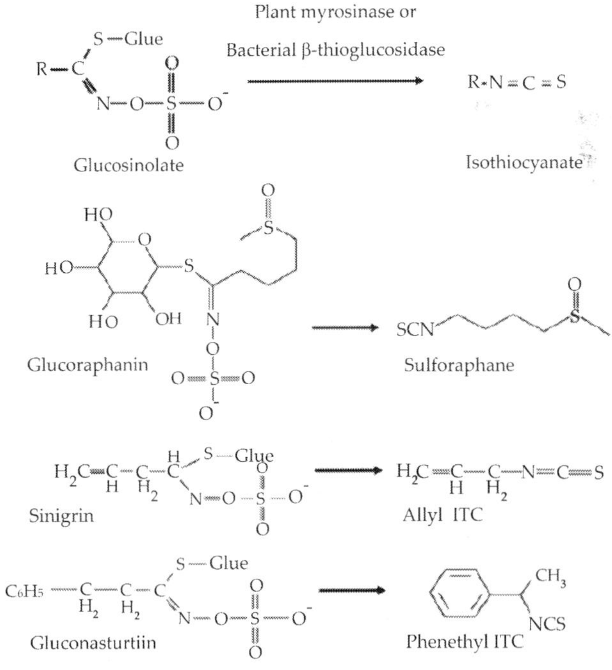

Figure 11: Conversion of select glucosinolates to their corresponding isothiocyanates.

Phenolic Acids

A phenolic acid is a type of polyphenol. Phenolic acids are found in a variety of plant-based foods; the seeds and skins of fruits and the leaves of vegetables contain the highest concentrations.

Phenolic acids are readily absorbed through the walls of the intestinal tract, and they may be beneficial to health because they work as antioxidants that prevent cellular damage due to free-radical oxidation reactions. They may also promote anti-inflammatory conditions in the body when eaten regularly.

Phenolic acids are generally abundant in a balanced diet. They should beplentiful as long as adequate amounts of fruits, vegetables, and whole grains are consumed. Examples of foods rich in phenolic acids include mangos, berries, apples, citrus fruits, plums, cherries, kiwis, onions, tea, coffee, red wine, and flour made from whole wheat, rice, corn or oats.

There are many different phenolic acids found in nature, and they can be divided into two categories:

- Benzoic acid derivatives, such as gallic acid
- Cinnamic acid derivatives, including caffeic acid and ferulic acid.

The cinnamic acids are more common than the benzoic acids.

Gallic acid is found in tea and grape seeds. Coffee contains caffeic acid and chlorogenic acid. Blueberries, kiwis, plums, cherries and apples contain large amounts of caffeic acid. Drinking coffee has been associated with a decreased risk of cancer and other chronic diseases. It also contains caffeine and other potentially beneficial compounds. Red wine has phenolic acids, but it also has resveratrol, which is another type of polyphenol that has potential health benefits.

Red wine and citrus fruits contain cinnamic acid. Ferulic acid is found in the outer coverings of cereal grains. Corn flour has the most ferulic acid, and whole grain wheat, rice, and oat flours are good sources of ferulic acid as well.

Phenolic acids may be available commercially in the form of dietary supplements, such as grape seed extract or green tea extract, which contain gallic acid. These supplements are usually marketed as antioxidants, but current research evidence suggests that fruits, vegetables, and whole grains are more beneficial than taking any antioxidant supplements.

Studies using cell and animal models show the effects of phenolic acids on both the expression and activity of enzymes involved in the production of inflammatory mediators. For example, COX-2 expression and activity are reduced by p-coumaric acid in dextran/sodium sulphate (DSS)-induced inflammation in a rodent model and caffeic acid suppresses the expression of IL-17 and ameliorates DSS-induced colitis in mice. Clearly, these compounds have anti-inflammatory properties in model systems but results have to be interpreted with caution in a nutritional context. In such studies doses often markedly exceed that which may be achievable from the diet. Whether analogous effects occur at dietary relevant concentrations is often unclear. Although some observational studies suggest inverse associations between the consumption of phenolic-rich diets and inflammatory markers, in general, the impact of dietary phenolic acids on gut health in human subjects has not yet been assessed in adequately powered and controlled dietary intervention trials. Many potentially beneficial effects of phenolic acids interpolated from in vitro studies may not be of nutritional relevance unless phenolic acids gain access in vivo to appropriate cellular sites such as the colonocytes lining the gut. Systemic and colonic bioavailability of phenolic acids is not well understood and the complexity of absorption and hepatic and microbial metabolism make the correlation of dietary intake with physiological effects distinctly problematical. The exact mechanisms for the absorption of phenolic acids are not clear, but may be passive or involve H + and Na + transport systems and/or monocarboxylic acid transporters. The small intestine and colon can both be absorption sites for phenolic acids, with conjugated forms mainly present in the diet generally having a low bioavailability compared with the aglycones. These observations strongly suggest that the bulk of the cinnamic acids escape early absorption in the gastro-intestinal tract. They appear to be destined for the colon where they are released and metabolised by the microbiota. Moreover, post-prandial concentrations of phenolic acids in plasma are low, indicative of low levels of absorption and/or rapid hepatic metabolism and excretion. It is likely that any bioactivity that contributes to gut health may predominantly be located in the lower gastro intestinal tract.

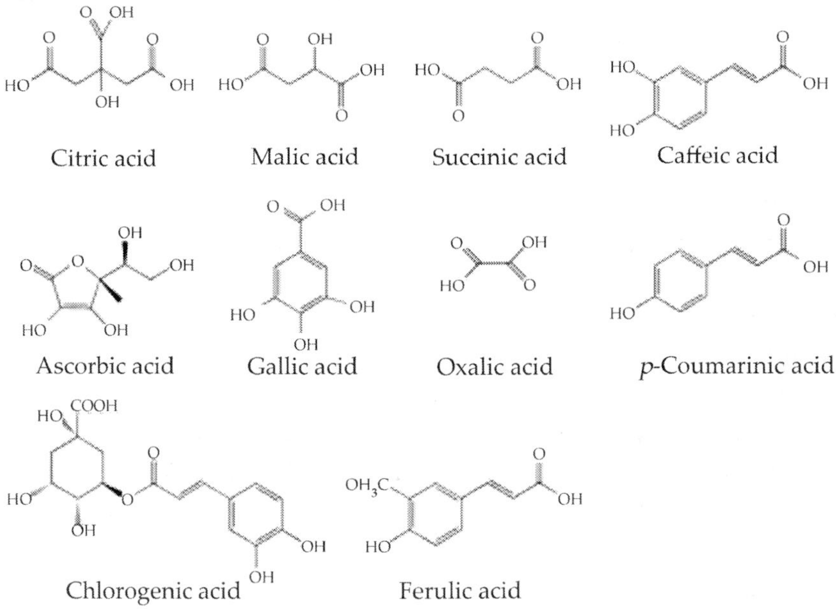

Figure 12: Chemical structure of phenolic acids.

The human gut is exposed to a plethora of simple phenolic acids and related metabolites arising directly from the food consumed and via the microbial degradation of more complex dietary phenolic compounds. Many of these metabolites have potent bioactivities in model systems, which could potentially contribute to gut health. However, at present it remains largely unknown which of these compounds are beneficial or indeed detrimental. Moreover, understanding of their mechanisms of action in vivo and how they interact with the gut microbiota is likely to help in the prevention of conditions such as inflammatory bowel disease and colon cancer. Future research directions are likely to expand on current metagenomic and metabolomic approaches to elucidate which pro-inflammatory and potentially carcinogenetic changes in gene expression can be moderated by simple phenolic acids. In addition, the development of good methods for monitoring total bacterial communities and their metabolic activity in response to phenolic acids is essential. At present, there is arguably insufficient information to allow the recommendation of specific plant-based foods rich in particular phenolics to optimise gut health. However, in view of the preventative effects of acetylsalicylic acid on the development of colon cancer, more studies on the nutritional efficacy of phenolic acid-rich foods appear warranted.

Sterols and Stanols

Although plants usually only contain a small amount of fat, their seeds are relatively concentrated sources. Interest in one particular group of plant-derived lipids, plant stanols and sterols, has developed in relation to their potential beneficial health effects, particularly in respect to cardiovascular disease. Sterols are essential components of cell membranes, with a key role in controlling membrane fluidity and permeability. They are present naturally in small quantities in many fruits, vegetables, nuts, seeds and legumes. Stanols are chemically similar to sterols. They occur in similar sources such as nuts, seeds and legumes but in smaller quantities than sterols. To improve their solubility, stanols and sterols are often combined with a fatty acid ester to produce plant stanol and plant sterol esters and it is in this form that is often added to food products.

The structure of plant stanols and plant sterols is very similar to that of cholesterol and so they are able to compete with cholesterol in the human gut. It is thought that including plant stanols and sterols in the diet reduces the absorption of cholesterol (both that provided by the diet and that which reaches the gut via bile acids produced in the liver that pass into the gut via the gall bladder). The unabsorbed cholesterol is excreted via the large bowel and this results in much less cholesterol reaching the blood supply and hence a lower blood cholesterol concentration. Total and LDL cholesterol levels are lowered without affecting HDL concentrations. Approximately 30-60% of total cholesterol is absorbed from the intestine into the blood; however, when plant stanol and sterol esters are present, absorption of cholesterol falls to approximately 20%. The reduction of LDL cholesterol ranges between 6 to 15%. Lowering blood total and LDL cholesterol levels may have the beneficial effect of reducing coronary heart disease risk.

In the context of the regulation on health claims, the European Food Safety Authority (EFSA) recently concluded that the level of cholesterol in the blood can be reduced, on average by 7 to 10.5%, if a person consumes 1.5 to 2.4 grams of plant sterols or plant stanols every day. EFSA found evidence that the effect is usually established within the first 2-3 weeks and can be sustained for at least 85 weeks.

EFSA also concluded that foods such as yoghurts and milk, including low-fat yoghurts and cheese, fat spreads, mayonnaise, salad dressing and other dairy products, were the most suitable for delivering the cholesterol-lowering effects from plant stanols and sterols to the body. For other foods, either information was lacking or they appeared to be less effective in reducing blood cholesterol levels.

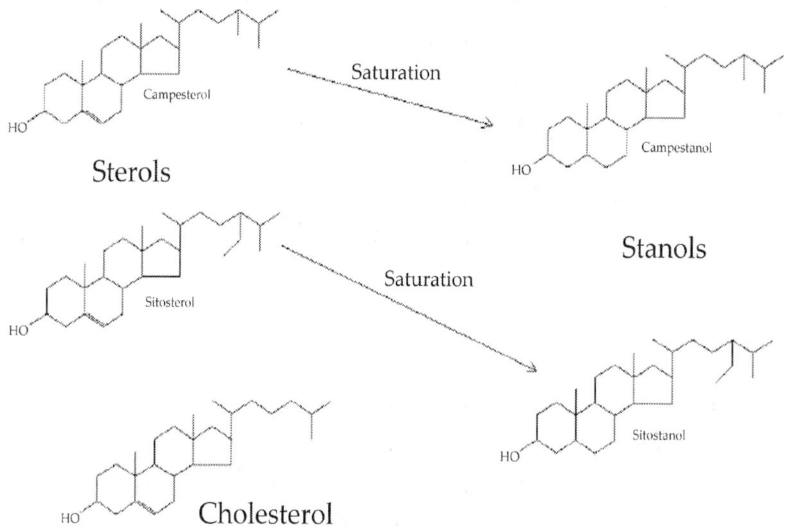

Figure 13: Plant sterols and stanols.

These days, products fortified with plant stanol or sterol esters are widely available. Therefore, when consumed regularly, and as part of a healthy, varied diet, foods and drinks with added plant stanol or sterol esters can bring about a significant reduction in LDL cholesterol concentration. Thus, they can be a useful addition to a healthy diet for individuals wishing to reduce their blood cholesterol levels.

The intake of plant sterols in a Western diet is 200–400 mg/d (Jones et al. 1997), and a vegetarian diet may provide twice that amount (Ling & Jones, 1995). Plant sterols are obtained from the diet by intestinal absorption. The absorption rate for sitosterol is about 5 %, but higher absorption rates of 10 % have been reported for campesterol (Salen et al. 1970; Heinemann et al. 1993). Plant sterols apparently inhibit the intestinal absorption of cholesterol, possibly by displacing cholesterol from micelles and thus reducing absorption (Ikeda et al. 1989). In earlier studies, very large doses (up to 50 g/d) of plant sterols were used to reduce cholesterolaemia (Best et al. 1955). Later, smaller doses (3–18 g/d) were reported to reduce serum total cholesterol and LDL-C by 10–20 % (Grundy & Mok, 1976; Schlierf et al. 1978; Becker et al. 1992, 1993). The maximum effect was obtained with 3 g plant sterols/d (Grundy & Mok, 1976; Lees & Lees, 1976; Lees et al. 1977). In addition, Pelletieret al. (1995) reported that plant sterol levels as low as 1 g/d can reduce cholesterolaemia. Plant sterols are poorly absorbed, and are believed to be free of adverse effects. High doses of plant sterols may reduce serum levels of fat-soluble vitamins (Gylling et al. 1996; Weststrate & Meijer, 1998), and very high doses of plant sterols may affect reproduction in animals (Malini &Vanithakumar, 1993; MacLatchy & Van der Kraak, 1995).

Polyols

Polyols are hydrogenated carbohydrates used as sugar replacers. Polyols, also called sugar alcohols, are a group of versatile, reduced-calorie carbohydrates that provide the taste and texture of sugar with about half the calories. They are used as food ingredients to replace sugar in an increasing variety of sugar-free and reduced-calorie foods and beverages for their functional and health benefits. These products include chewing gums, candies, ice cream, baked goods and fruit spreads. In addition, they function well in fillings and frostings, canned fruits, beverages, yogurt and tabletop sweeteners. Interest now arises because of their multiple potential health benefits. They are claimed to be non-cariogenic (sugar-free tooth-friendly), low-glycaemic (potentially helpful in diabetes and cardiovascular disease), low-energy and low-insulinemic (potentially helpful in obesity), low-digestible (potentially helpful in the colon), osmotic (colon-hydrating, laxative and purifying) carbohydrates. These attributes, however, remain under study.

Polyols are widely found in foods, for example apricots, mushrooms, cauliflower; and also commonly used as artificial sweeteners in sugar free chewing gum, lollies and mints. In the food industry polyols are often identifiable by additive numbers on food packaging: Sorbitol (420), Mannitol (421), Maltitol (965), Xylitol (967) and Isomalt (953), where they are usually present along with the mandatory declaration "excessive consumption may have a laxative effect." Those currently used in foods in the U.S. are erythritol, hydrogenated starch hydrolysates (including maltitol syrups), isomalt, lactitol, maltitol, mannitol, sorbitol and xylitol. Polyols, are commonly added to foods because of their lower caloric content than sugars as listed onTable 8 below.

Table 8: Caloric value of polyols

Sweetener	Calories per Gram
Sucrose	4
Polyglycitols	3
Sorbitol	2.6
Xylitol	2.4
Maltitol	2.1
Isomalt	2
Lactitol	2
Mannitol	1.6
Erythritol	0.2

Polyols do not have associated transport systems in the small intestine and are thought to be absorbed in the epithelium by passive diffusion. Their diffusion depends on:

- the molecular size of the polyols
- variation of pore size
- transit time

Pore size can also be affected by mucosal disease and is often reduced in coeliac disease. Limited studies reviewing absorption of sorbitol and mannitol have produced considerable variation and its availability for fermentation varies with dose taken.

Polyols have been used in foods around the world for many years as sugar replacers. An Expert Committee of the World Health Organization has carefully reviewed them and concluded that they are safe for human consumption. In the U.S., the FDA classifies some as GRAS and others are approved food additives.For the vast majority of consumers, these sweeteners do not cause a problem. In some people, excessive consumption may cause gastrointestinal symptoms, such as gas or laxative effects, similar to reactions to beans and certain high-fibre foods. Such symptoms depend on an individual's sensitivity and the other foods eaten at the same time.

Prebiotics

The term prebiotic was first used in 1995 and can be defined as a non-digestible food ingredient that can deliver beneficial effects on health by selectively stimulating the growth and/or activity of specific health-promoting bacteria in the colon. In other words, prebiotics promote the growth of particular bacteria in the gut that are beneficial to intestinal health (for example Lactobacillus sp., Bifidobacteria sp. and Lactococcus sp.). They also inhibit the growth of bacteria that are potentially harmful to intestinal health (for example bacteria that produce toxins such as Clostridia and Escherichia coli). This is different from probiotics, which are live bacteria capable of journeying through the gut into the large bowel. By helping increase the amount of 'good' bacteria, prebiotics are thought to promote a healthy environment in the intestine and so could potentially have health benefits.

The main types of prebiotics are fructo-oligosaccharides, e.g. inulin, and oligofructose and lacto-oligosaccharides, e.g. lactulose, galacto-oligosaccharides and lactosucrose. Foods naturally containing substances with prebiotic properties include leeks, chicory, asparagus, bananas, artichokes, garlic, onion, wheat, soybean and oats, and more recently some honeys have been suggested to have prebiotic properties. Sometimes foods with prebiotic properties are added to foods and are referred to as functional foods. Prebiotic-containing foods can be found in supermarkets and include, for example, some cheese products, yogurts and yogurt drinks, breakfast cereals and cereal bars.

Regular consumption of prebiotics has been suggested to have a wide range of potential health benefits including aiding mineral absorption (e.g. calcium), improving immune function, reducing blood cholesterol levels, playing a role in cancer prevention and helping to relieve constipation and irritable bowel syndrome. However, the strength of the evidence is stronger for some of these claims than others and further research in humans is required to determine whether some of these benefits exist when prebiotic-containing foods are consumed as part of a normal diet.

To date, the most promising areas are calcium absorption and immune system effects.

There is a growing body of evidence demonstrating a positive effect of prebiotics on the intestinal flora in infants by increasing the number of 'good' bacteria in the gut. Accumulating research suggests that prebiotic infant formulas may help to alleviate conditions such as atopic dermatitis (allergic skin condition), constipation and to reduce the risk of infection in early life.

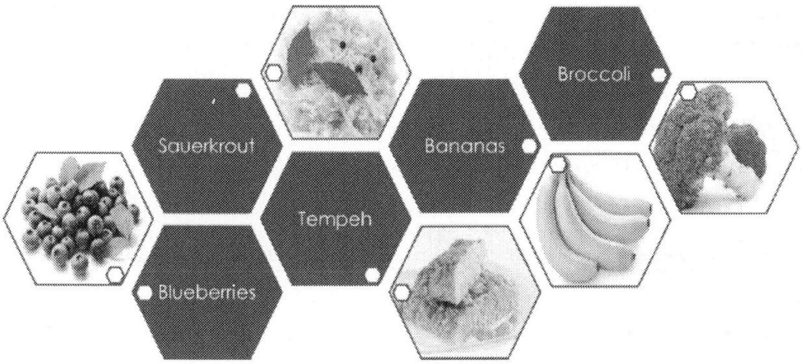

Figure 14: Food containing naturally occurring prebiotics to aid in the restoration of gut microbiota.

Figure 15: Probiotic-rich foods.

Phytoestrogens

The word phytoestrogens comes from the Greek word "phyto," or plant, and "estrogen," the hormone that causes fertility in all female mammals.

Phytoestrogens have also been termed dietary estrogens because they are not created by the human endocrine system. They can only be ingested through the diet or dietary supplements.

A similar class of non-endocrine estrogens is xenoestrogens, synthetic estrogens found in certain kinds of plastic and pesticide products. It is important to consider the combination and interaction of all of the environmental estrogens routinely encounter to truly evaluate their possible influence on health.

In their natural state, phytoestrogens exist within plants as a natural defence against herbivores. Plants secrete these hormones to modulate the fertility of animals that may eat them to reduce further attacks (Hughes, 1988). Soy is known as the most phytoestrogen-rich plant found in a typical Western diet.

Part of what makes phytoestrogens of high research interest is their ability to both mimic estrogen and act as an estrogen antagonist (meaning they behave in the opposite way of biological estrogen). They affect the body by attaching to estrogen receptors. Because up until now they are not considered specifically necessary for the human diet, phytoestrogens cannot be considered actual nutrients. As the scientific understanding of how they work and interact with other nutrients, this might change. The most well-studied of the types of phytoestrogen compounds are isoflavones, also commonly referred to as soy isoflavones because most are found in soy and red clover.

The estrogenic and anti-estrogenic effects of phytoestrogens have often been thought of as overwhelmingly negative. For the majority of young women, extra estrogen in the body can lead to infertility, polycystic ovarian syndrome and even certain types of cancer. Men generally don't need extra estrogen in their systems, either. However, in the case of women over the age of 50 experiencing decreased estrogen levels, extra estrogen can decrease cancer risk, among other benefits.

The question of whether or not phytoestrogens are beneficial or harmful to human health remains unresolved. The answer is likely complex and may depend on age, health status, and even the presence or absence of specific gut microflora. Clarity on this issue is needed because global consumption is rapidly increasing through both natural sources and dietary supplements. Phytoestrogens are present in numerous dietary supplements and widely marketed as a natural alternative to estrogen replacement therapy. Soy infant formula now constitutes up to a third of the US market, and soy protein is now added to many processed foods. As an estrogen agonists/antagonists with molecular and cellular properties similar to synthetic endocrine disruptors such as Bisphenol A (BPA), phytoestrogens provide a useful model to comprehensively investigate the biological impact of endocrine disruptors in general (Patisaul and Jefferson, 2010).

Accumulating evidence from molecular and cellular biology experiments, animal studies, and, to a limited extent, human clinical trials suggests that phytoestrogens may potentially confer health benefits related to cardiovascular diseases, cancer, osteoporosis, and menopausal symptoms. These potential health benefits are consistent with the epidemiological evidence that rates of heart disease, various cancers, osteoporotic fractures, and menopausal symptoms are more favourable among populations that consume plant-based diets, particularly among cultures with diets that are traditionally high in soy products.

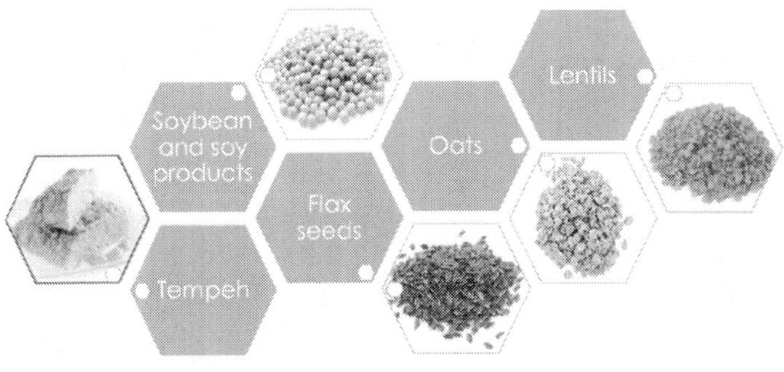

Figure 16: Top phytoestrogen containing foods.

Table 9: Classification, dietary sources and structure of common phytoestrogens

Group	Subgroup	Examples	Dietary sources	**Basic structure**
17β-Estradiol	Endogenous Estrogen	N/A	N/A	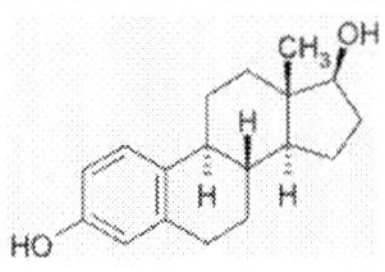
Polyphenols		Resveratrol	Grape skin, red wine	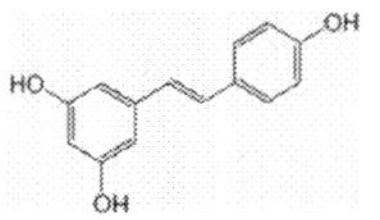
Flavonoids	Flavanones	Eriodictyol, Hesperetin, Homoeri-odictyol, Naringenin	Citrus fruits and juices	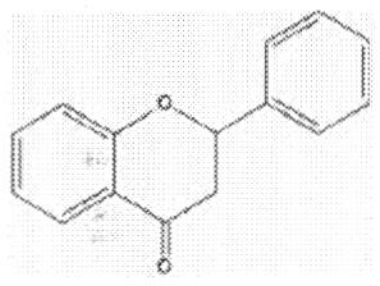
	Flavones	Apigenin, Luteolin, Tangeritin	Parsley, celery, capsicum pepper	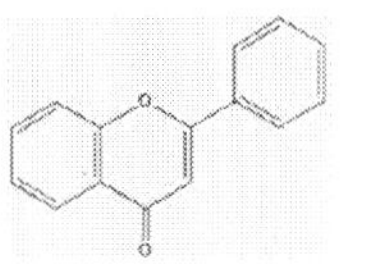
	Flavonols	Fisetin, Kaempferol, Myricetin, Pachypodol, Quercetin, Rhamnazin	Kale, broccoli, onions tomatoes, lettuce, apples, grapes, red wine	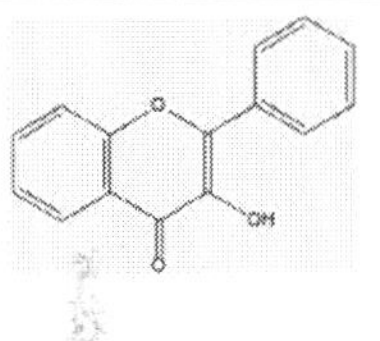
	Catechins	Proantho-cyanides	Chocolate, green tea, beans, apricots, cherries, berries	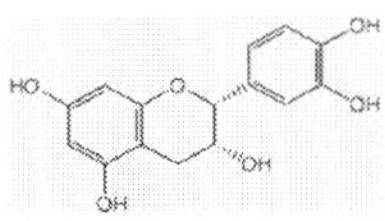

The structure of 17β-estradiol is provided as a reference for comparison

While research supports these benefits described above, there are negative effects of phytoestrogens as well, of which consumers should be aware. These issues are mainly related to fertility and development. It is also important to note that many of these studies delve specifically into the effects of phytoestrogens in soy, which contains its own problematic issues. Research has found that regular estrogen intake:

- Can adversely affect fertility: Studies show that diets rich in certain phytoestrogens reduce the occurrence of pregnancies in humans, California quail, deer mice, Australian sheep and cheetahs. In some of these examples, the removal of phytoestrogens from the diet caused fertility levels to rebalance. In addition, exposure to genistein and coumestrol, two specific phytoestrogen compounds, early in life may contribute to fertility issues later in development. They may cause decreased sperm levels, but this remains unclear throughout different research studies. (Mumford et al., 2015 and Mitchell et al., 2001)

- May lead to hormonal issues: One particular area of concern is the appearance of plant-based estrogens in soy infant formulas, as the long-term consequences are still not well-understood. However, some research indicates possible negative correlations, including a higher occurrence of boys born with hypospadias, increased use of allergy medicines, and more severe menstrual bleeding and cramps for girls. Again, genistein was identified as the most likely culprit. As with much of the information about phytoestrogens, there are conflicting ideas about whether or not they're harmful in the long run. At least one study found no significant difference in people fed soy formula versus those fed with cow's milk, so more research is still needed to make any firm conclusions. (Miniello et al. 2003)
- Potentially stimulate breast cancer growth: As already mentioned, phytoestrogens have the ability to inhibit the growth of certain cancers, such as breast cancer. However, a study out of Canada found that low concentrations of phytoestrogens actually may make breast cancers grow faster, as well as inhibit the effects of tamoxifen, the drug used to treat late-stage breast cancer. In higher concentrations, the effect was the opposite, causing tumours to shrink and magnifying the impact of the drugs — showing just how complicated phytoestrogen activity can be (de Lemos, 2001).
- Increase risk of cognitive decline and dementia: Another concern to be aware of is the possibility of mental decline in

connection with consuming large amounts of phytoestrogens. While the evidence remains inconclusive, research suggests a link between dementia and cognitive decline with phytoestrogen intake (Soni et al., 2016).

Phytoestrogens exist in many foods, supplements and essential oils. A wide range of commonly consumed foods contain appreciable amounts of these different phytoestrogens. Some of the highest concentrations can be found in (Zava et al, 1998):

- Soybeans and soy products
- Tempeh
- Flaxseeds
- Oats
- Barley
- Lentils
- Sesame seeds
- Yams
- Alfalfa
- Apples
- Carrots
- Jasmine oil
- Pomegranates
- Wheat germ
- Coffee
- Licorice root
- Hops
- Bourbon
- Beer
- Red clover
- Clary sage oil

Table 10: Examples of functional components for the development of functional foods

Class/Components	Source	Potential Benefit	Tips to Include Healthful Components in the Diet
CAROTENOIDS **Beta-Carotene**	Carrots, pumpkin, sweet potatoes, cantaloupe, spinach, tomatoes	Neutralizes free radicals which may damage cells; bolsters cellular antioxidant defenses; can be made into Vitamin A in the body	For Beta-carotene rich "fries", thinly slice sweet potatoes, lightly coat with olive oil or fat-free cooking spray, add spices (pepper, rosemary, thyme) to taste and bake in 425-degree oven until golden brown (10-15 mins). Time-saver: Buy pre-cut sweet potatoes in the frozen foods section.
Lutein, Zeaxanthin	Kale, collards, spinach, corn, eggs, citrus fruits, asparagus, carrots, broccoli	Supports maintenance of eye health	Enjoy a crisp spinach with hard-boiled egg slices For a lutein-rich breakfast, make a spinach omelet. Beat 2 eggs; stir in ¼ C milk and ½ C spinach (fresh or frozen and drained). Lightly coat skillet with fat-free cooking spray. Cook on low heat until set. Kale, which provides the same health benefits, can be an easy substitute for spinach
Lycopene	Tomatoes and processed tomato products, watermelon, red/pink grapefruit	Supports maintenance of prostate health	Like other carotenoids, lycopene is best absorbed from a meal containing some oil. If you love tomatoes, try adding tomato sauce to sautéed zucchini for a fun and colorful side dish! Sprinkle a little sugar or low-calorie sweetener on sliced grapefruit before eating to bring out the natural sweetness within.

Dietary (Functional and Total) Fiber

Insoluble fiber	Wheat bran, corn bran, fruit skins	Supports maintenance of digestive health; may reduce the risk of some types of cancer	Add a little dry wheat bran when making smoothies or muffins to bulk up the fiber content
Beta glucan**	Oat, brad, oatmeal, oat flour, barley, rye	May reduce risk of coronary heart disease (CHD)	Jazz up your oatmeal with these tasty extras: 1 tbsp. peanut butter and jelly; a sprinkle of cinnamon and pumpkin puree; silvered almonds and ½ C frozen berries. Keep instant oatmeal packets in your desk drawer to have on hand when you miss breakfast or need a hearty afternoon snack to tide you over until dinner.
Soluble fiber **	Psyllium seed husk, peas, beans, apples, citrus fruits	May reduce risk of CHD and some types of cancer	Try adding canned beans (black, pinto or garbanzo) to your regular meal – layer them in quesadillas, add to an omelet or enjoy in a mixed green salad
Whole grains**	Cereal grains, whole wheat bread, oatmeal, brown rice	May reduce risk of CHD and some types of cancers; supports maintenance of healthy blood glucose levels	Keep a few mini-bags of microwave popcorn in your desk drawer to pop for a quick, low-calorie snack. Spice up your popcorn with garlic powder and cinnamon or rosemary and parmesan cheese. Yum! Are your kids picky eaters? Make their next sandwich visually appealing: use one slice of white bread and one slice of whole grain bread, cut sandwich into four pieces, turn two sections to create a checker board effect

*Examples are not an all-inclusive list

**FDA-approved health claim for component

Class/Components	Source*	Potential Benefit	Tips to Include Healthful Components in the Diet
FATTY ACIDS **Monounsaturated fatty acids (MUFAs)****	Tree nuts, olive oil, canola oil	May reduce risk of CHD	Make portion-controlled snack bags of mixed nuts such as almonds, pecans and cashews and add dried fruit for extra flavor and fiber. Keep the bags on hand to grab on your way out of the door for a quick and easy mid-morning or afternoon snack
Polyunsaturated fatty acids (PUFAs) – **Omega -3 fatty acids - ALA**	Walnuts, flaxseeds, flaxseed oil	Supports maintenance of heart and eye health; supports maintenance of mental function	When cooking, substitute flax seed oil for canola or olive oil one to two times per week. Keep in mind, flax seed oil is fragile; refrigerate in a dark container and use at low-temperature for stir-frying.
PUFAs- Omega- 3 fatty acids – DHA/EPA**	Salmon, tuna, marine and other fish oils	May reduce risk of CHD; supports maintenance of eye health and mental function	Salmon or tuna canned in water or a shelf-stable pouch can make easy and affordable meals; add a few tablespoons of olive oil, season with pepper and thyme and enjoy on top of whole grain crackers or wrapped with tomatoes in a lettuce leaf.
Conjugated linoleic acid (CLA)	Beef and lamb; some cheese	Supports maintenance of desirable body composition and immune health	Bring Hawaii to your next cookout by preparing these kebabs for the grill; alternate pineapple, beef, onions and bell peppers on wooden skewers; season with garlic powder, pepper, paprika, oil and lime.
FLAVONOIDS **Anthocyanins – Cyanidin, Pelargonidin, Delphinidin, Malvidin**	Berries, cherries, red grapes	Bolster cellular antioxidant defenses; supports maintenance of healthy brain function	Add berries to your yogurt or morning oatmeal for extra flavor. For a cold treat, enjoy a smoothie! Combine 1 C frozen berries, ½ banana, ¾ C Yogurt, and ¾ C milk in a blender; blend until smooth.

Flavanols – Catechins, Epicatechins, Epigallocatechin	Tea, cocoa, chocolate, apples, grapes	Supports maintenance of heart health	Make your own iced tea by pouring boiling water over 3-4 bags of your favorite tea; let sit for 5 minutes, then pour over pitcher of ice and fill to the brim with water; stir and refrigerate to chill. Enjoy with fresh-squeezed lemon
Procyanidins and Proanthocyanidins	Cranberries, cocoa, apples, strawberries, grapes, red wine, peanuts, cinnamon, tea, chocolate	Supports maintenance of urinary tract health and heart health	Dice apples and simmer in water until soft, add to blender with dash of cinnamon; puree. Enjoy as a spread on whole wheat toast or crackers!
Flavanones – Hesperetin Naringenin	Citrus fruits	Neutralizes free radicals which may damage cells; bolster cellular antioxidant defenses	Squeeze ½ orange and ½ lemon into a small dish; add olive or flax oil, dashes of salt, pepper and basil for a perfectly refreshing salad dressing. Rushed for time? Get the same benefits from orange or grapefruit juice.
Flavonols – Quercetin, Kaempferol, Isorhanetin, Myricetin	Onions, apples, tea, broccoli	Neutralizes free radicals which may damage cells; bolster cellular antioxidant defenses	Caramelized onions make a sweet and tasty garnish to many main dishes. Sauté onions over low heat in 1 tsp. oil until deep gold in color; use to pop prepared steak, chicken, or fish.

*Examples are not an all-inclusive list

**FDA-approved health claim for component

Class/Components	Source*	Potential Benefit	Tips to include Healthful Components in the Diet
ISOTHIOCYANATES **Sulforaphane**	Cauliflower, broccoli, broccoli sprouts, cabbage, kale, horseradish	May enhance detoxification of undesirable compounds; bolsters cellular antioxidant defenses	Keep frozen broccoli and cauliflower on hand for an easy side dish. In a microwaveable bowl, cover frozen vegetables with paper towel and cook 1-1/2 minutes. Add silvered almonds and sprinkle with parmesan cheese, cook 30-60 seconds more or until the cheese melts
MINERALS **Calcium****	Sardines, spinach, yogurt, low-fat dairy products, fortified foods and beverages	May reduce the risk of osteoporosis	Add cooked spinach, canned artichokes and plain yogurt to blender or food processor; puree for a tasty cracker spread or dip. Season to taste – try black pepper, basil and garlic powder.
Magnesium	Spinach, pumpkin seeds, whole grain breads and cereals, halibut, almonds brazil, nuts, beans	Supports maintenance of normal muscle and nerve function, immune health and bone health	For extra flavor, sprinkle fish fillet with pumpkin seeds before baking or add them to veggie stir-fry
Potassium**	Potatoes, low-fat dairy products, whole grain breads and cereals, citrus juices, beans, banana, leafy greens	May reduce the risk of high blood pressure and stroke, in combination with a low sodium diet	Puree ripe bananas with crushed peanuts and portion into freezer cups to enjoy as frozen treat! For extra indulgence, add some fresh chocolate shavings.

Selenium	Fish, red meat, whole grains, garlic, liver, eggs	Neutralizes free radicals which made damage cells; supports maintenance of immune and prostate health	Roasting garlic gives it a rich, buttery flavor. Cut a full bulb of garlic crosswise to expose wrapped cloves; drizzled with oil/flax oil, season with salt/pepper, cover with aluminum foil to bake in 400-degree oven for 30 minutes or microwave on high (no foil) for 1 minute, turn bulb upside down and cook 1 minute more. Delicious!
PHENOLIC ACIDS **Caffeic acid, Ferulic acid**	Apples, pears, citrus fruits, some vegetables, whole grains, coffee	Bolsters cellular antioxidant defenses; supports maintenance of eye and heart health	Love your morning coffee? Good news – coffee is a powerful source of antioxidants. Pair it with apples and peanut butter for an antioxidant-packed breakfast
PLANT STANOLS/STEROLS **Free Stanols/Sterols****	Corn, soy, wheat, fortified foods and beverages	May reduce risk of CHD	Get your free stanols/sterols from fortified foods such as bread containing "whole wheat flour", low-fat yogurt, soy milks and some cereals. Dietary supplements in soft gel form that contain free stanols and sterols are also available and can provide similar benefits when used as part of a low-fat, low-cholesterol diet

Stanol/Sterols esters**	Stanol ester dietary supplements, fortified foods and beverages, including table spreads	May reduce risk of CHD	Many table spreads (butter or margarine alternatives) are fortified with stanol and/or sterol esters. Other commercial products fortified with stanols and sterols include some orange juices, yogurt beverages, chocolate and granola bars. Check product labels and ingredients list to confirm the presence and amount of stanols and sterols

*Examples are not an all-inclusive list

**FDA-approved health claim for component

Class/Components	**Source***	**Potential Benefit**	**Tips to include Healthful Components in the Diet**
POLYOLS **Sugar alcohols** - Xylitol, Sorbitol, Mannitol, Lactitol**	Some chewing gums and other food applications	May reduce risk of dental caries	Reduce risk for dental caries and help curb your appetite by chewing sugar-free gum after eating
PREBIOTICS **Inulin, Fructoligosaccharides (FOS), Polydextrose**	Whole grains, onions, some fruits, garlic, honey, leeks, banana, fortified foods and beverages	Supports maintenance of digestive health; supports calcium absorption	Food products fortified with prebiotics are beginning to appear, but did you know that you can also get prebiotics by simply adding honey to some of your routine meals? Add honey to oatmeal or yogurt, drizzle it over a banana, spread it on your morning toast, or treat yourself to ½ peanut butter and honey sandwich.

PROBIOTICS **Yeast, Lactobacilli, Bifidobacteria and other specific strains of beneficial bacteria**	Certain yogurts and other cultured dairy and non-dairy applications	Supports maintenance of digestive and immune health; benefits are strain-specific	Add probiotics to your diet by choosing low-fat yogurt with a beneficial probiotic strain: choose a variety of flavored or plain yogurts with probiotics for a quick and tasty snack. Blend 6-8oz yogurt, 6oz orange juice, 1 tbsp. honey, ½ banana and ice for a 30-second smoothie. For a healthful breakfast, top pancakes with flavored yogurt and fresh fruit like strawberries and bananas
PHYTOESTROGENS **Isoflavones – Daidzein, Genistein**	Soybeans and soy-based foods	Supports maintenance of bone and immune health, and healthy brain function; for women, supports menopausal health	Try including tofu in some of your favorite recipes. Get your isoflavones by choosing soft, silken tofu and adding it to the cheese sauce mixture when making lasagna. Other recipes that work well with tofu additions include quiches, veggie dips, quesadillas, chili, chocolate mousse and even baked goods!
Lignans	Flax seeds, rye, some vegetables, seeds and nuts, lentils, triticale, broccoli, cauliflower, carrot	Support maintenance of heart and immune health	Add whole or ground flax seeds to a smoothie or baked goods recipe. You can also try occasionally substituting rye bread for your usual bread of choice at home or when eating at a restaurant.

SOY PROTEIN **Soy Protein****	Soybeans and soy-based foods like milk, yogurt, cheese and tofu	May reduce risk of CHD	Soybeans are also called "edamame" in many Asian cuisine restaurants; try ordering a plate to share before your meal arrives. When cooking at home, look for frozen edamame to easily prepare as a healthy snack or party sampler. Soy milks are another option for increasing soy protein intake. Check the ingredients of nutrition bars or meal replacements for "soy protein"

*Examples are not an all-inclusive list

**FDA-approved health claim for component

Class/Components	**Source***	**Potential Benefit**	**Tips to include Healthful Components in the Diet**
SULFIDES/THIOLS **Diallyl sulfide, Allyl methyl trisulfide**	Garlic, onions, leeks, scallions	May enhance detoxification of undesirable compounds; supports maintenance of heart, immune and digestive health	Scallions, or "green onions", are milden than traditional onions and are commonly used to garnish salads or cooked sauces. Incorporate scallions into meals by thinly chopping the green stalks, and adding them to mashed potatoes, cold pasta, salad, tuna salad or canned soups. Leeks can also be an easy substitute but are more commonly used in soups.
Dithiolthiones	Cruciferous vegetables	May enhance detoxification of undesirable compounds; supports maintenance of healthy immune function	Broccoli and cauliflower are the more commonly known cruciferous vegetables that we eat, but did you know that varieties of cabbage, bok choy, Brussel sprouts, kale and wasabi, among others, are also sources of dithiolthiones? Bok choy is great in any stir-fry or by itself. To cook: separate the white stem from the leaves and chop into 1" thick pieces. Add to oil in skillet on medium-high heat, sauté 3 minutes, add the leaves, and ½ cup of water or vegetable stock, stir until the leaves are wilted and cooked (about 5-10 minutes.) Season to taste and enjoy!

VITAMINS A***	Organ meats, milk, eggs, carrots, sweet potato, spinach	Supports maintenance of eye, immune and bone health; contributes to cell integrity	If you are a fan of boiled or raw carrots, try buying the larger carrots and slice them lengthwise into wide strips. Add them to the grill after lightly brushing with oil and season with fajita seasoning or other spices to give them a unique BBQ Flavor.
Thiamin (Vitamin B1)	Lentils, peas, brown or enriched white rice, pistachios and certain fortified breakfast cereals	Supports maintenance of mental function; helps regulate metabolism	If you are still unsure about making the transition to long-grain brown rice, start slow and create a half and half mixture of brown and white rice. Think about adding dried cranberries and silvered almonds to the rice mix as well for appealing color and a little sweetness to complement the dense "nutty" flavored rice.
Riboflavin (Vitamin B2)	Lean meats, eggs, green leafy vegetables, dairy products and certain fortified breakfast cereals	Supports cell growth; helps regulate metabolism	Get lots of vitamin B2 by bulking up your whole-wheat sandwich with lean-cut deli meat and double-stacked green and red leaf lettuce or spinach. The extra greens will give your sandwich that mouthwatering crunch in every bite.
Niacin (Vitamin B3)	Dairy products, poultry, fish, nuts, eggs and certain fortified breakfast cereals	Supports cell growth; helps regulate metabolism	For a spicy twist on chicken, grill chicken breasts and cut into strips, toss in your favorite hot sauce and serve with celery – healthful "chicken wings" for the next football party!
Pantothenic Acid (Vitamin B5)	Sweet potato, organ means, lobster, soybeans, lentils and certain fortified breakfast cereals	Helps regulate metabolism and hormone synthesis	Looking for a healthful crunch? Try dried soy nuts and dried fruit for a fiber-rich, vitamin B5-friendly snack. Soy nuts are easily added to yogurts, oatmeal, salads, stir-fries, chicken bakes and much more.

*Examples are not an all-inclusive list

**FDA-approved health claim for component

***Preformed vitamin A is found in foods that come from animals. Provitamin A carotenoids are found in many darkly colored fruits and vegetables and are major source of Vitamin A for vegetarians.

NATURALLY OCCURRING FUNCTIONAL FOODS

The majority of people believe that foods have specific benefits (IFIC, 2013) and can be used to prevent, manage and treat different health conditions. Further, food can also be used to reach and stay at optimal health. Conventionally, food healthiness has been associated with nutritional factors such as fat, fibre, salt and vitamin content. In addition to this conventional or traditional healthiness, food may contain other components that may have a positive impact on our well–being. Products that are claimed to have special beneficial physiological effects in the body have been called nutraceuticals, pharma foods, designer foods, nutritional foods, medical foods or super foods. More usually they are named as functional foods. The natural properties of food that may qualify them as a functional food can include, for example, the phytonutrient components in fruits and vegetables, whole grains and fibre in certain breads and cereals, as well as the calcium in milk. These properties are considered functional as they go beyond the provision of energy and micronutrients.

Plant Sources of Functional Foods

Grains and legumes

Dietary carbohydrates may influence the development of type 2 (non-insulin-dependent) diabetes, for example, through effects on blood glucose and insulin concentrations. Data support a protective role for grains (particularly whole grains), cereal fibre, and dietary magnesium in the development of several chronic health conditions. For example, whole grain consumption is inversely associated with body mass index in school children even after adjustment for body weight potential dietary effectors (fruit, vegetable and dairy intakes) (Choumenkovitch et al., 2013). Whole grains consumption may also be protective against stroke and metabolic syndrome.

Whole grains contain antioxidants, dietary fibre (inulin, beta-glucan, resistant starch), phenolics, phytate, phyto-estrogens, phytostanols, phytosterols, vitamins (carotenoids, tocotrienols, tocopherols) and minerals. Fiber-rich cereals consumption may help reduce cholesterol, hypertension, constipation and obesity. Rice protein improves mammals' body weight and lipids level because they have lower digestibility, which help upregulate lipolysis and downregulate lipogenesis (Yang et al., 2012). Nuts (tree nuts and peanuts) are nutrient dense (rich in unsaturated fats, protein, fibre, minerals, tocopherols, phytosterols, phenolics and other bioactive compounds) but epidemiologic and intervention studies showed they do not contribute to weight gain. Nut consumption showed reduced coronary heart disease and gallstones incidences in both genders and diabetes in women (Ros, 2010). Nuts have antioxidant, hypocholesterolemic, cardioprotective, anticancer, anti-inflammatory, anti-obesity and antidiabetic properties (Vadivel, Kunyanga, & Biesalski, 2012). Sesame contains sesamol, the powerful antioxidant that alleviates diet-induced inflammation and cardiometabolic syndrome in mammals by up-regulating PPARg, PPAR a and e-NOS. Sesamol dose-dependently decreased TBARS, nitrotyrosine, insulin resistance, hyperinsulinemia, hyperglycaemia, dyslipidaemia, TNF-alpha, IL-6, leptin, resistin, highly sensitive C-reactive protein (hs-CRP), hepatic transaminases and alkaline phosphatase. Sesamol normalizes adiponectin, nitric oxide, arterial pressures, antioxidant enzyme activities, hepatic steatosis and hepatocytes deterioration. Sesamol decreases LXRa (a nuclear receptor protein), SERBP-1c, P-JNK and NF-KB expression (Sharma et al., 2012). Early-life soya intake produced higher leptin and MCP-1 levels but soya intervention did not affect inflammation markers in men (Maskarinec, Oum, Chaptman, &Ognjanovic, 2009). In animals, isoflavones (e.g genistein) lowered plasma glucose, liver damage markers (aspartate amino transferase, alanine amino transferase). Isoflavones had significant effect on plasma insulin, leptin, blood sugar, triglycerides and glucagon in lean rats but not in obese rats (Ali, Velasquez, Hansen, Mohamed, & Bhathena, 2005).

The polyphenol-rich black soybean seed coat extract containing cyanidin 3-glucoside, catechins, and procyanidins amongst others, suppressed abdominal fat accumulation, plasma glucose level, major inflammatory cytokines, tumour necrosis factor-a and monocyte chemoattractant protein-1. The extract enhanced insulin sensitivity, UCP-1 and UCP-2 expression in the high-fat diet mice, to deter obesity and diabetes by enhancing energy expenditure and suppressing inflammation (Kanamoto et al., 2011).

Many legumes (e.g Phaseolus vulgaris) inhibit alphaamylase (Obiro, Zhang, & Jiang, 2009). Pigeon pea (Cajanus cajan L.) has linoleic acid and phytosterol (B-sitosterol, campesterol, stigmasterol) that synergistically produce antioxidant and antidyslipidemic effects by promoting cholesterol conversion to bile acid and increasing hepatic carnitine palmitoyltransferase-1 (CPT-1), LDL receptor, and cholesterol 7 a-hydroxylase (or cytochrome P450 7A1, CYP7A1) expression in high fat diet-fed hamsters (Dai et al., 2013). Chia seed in contrast does not promote weight loss or alter disease risk factors in overweight adults (Nieman et al., 2009).

Oats: Oat products are a widely studied dietary source of the cholesterol-lowering soluble fibre b-glucan. There is now significant scientific agreement that consumption of this particular plant food can reduce total and LDL cholesterol, thereby reducing the risk of CHD. For this, the FDA awarded the first food-specific health claim in January 1997 in response to a petition submitted by the Quaker Oats Company (Hasler, 1998). In its health claim petition, the Quaker Oats Company summarized 37 human clinical intervention trials con-ducted between 1980 and 1995. The majority of these studies revealed statistically significant reductions in total and LDL-cholesterol in hypercholesterolemic subjects consuming either a typical American diet or a low-fat diet. The daily amount of oat bran or oatmeal consumed in the above studies ranged from 34 g to 123 g. Quaker Oats determined that 3 g of b-glucan would be required to achieve a 5% reduction in serum cholesterol, an amount equivalent to approximately 60 g of oatmeal or 40 g of oat bran (dry weight). Thus, a food bearing the health claim must contain 13 g of oat bran or 20 g oatmeal, and provide, without fortification, at least 1.0 g of b-glucan per serving. In February of 1998, the soluble fibre health claim was extended to include psyllium fibre.

Soy:Soy has been in the spotlight during the 1990s. Not only is soy a high-quality protein, as assessed by the FDA's "Protein Digestibility Corrected Amino Acid Score" method, it is now thought to play preventive and therapeutic roles in cardiovascular disease, cancer, osteoporosis, and the alleviation of menopausal symptoms.

The cholesterol-lowering effect of soy is perhaps the most well-documented physiological effect related to functional foods apart from the digestive benefits of dairy products. A 1995 meta-analysis of 38 separate studies (involving 743 subjects) found that the consumption of soy protein resulted in significant reductions in total cholesterol (9.3%), LDL cholesterol (12.9%), and triglycerides (10.5%), with a small but insignificant increase (2.4%) in high density lipoprotein (HDL) cholesterol (Anderson et al., 1995). Linear regression analysis indicated that the threshold level of soy intake at which the effects on blood lipids became significant was 25g.

Regarding the specific component responsible for the cholesterol-lowering effect of soy, recent attention has focused on the isoflavones (Potter, 1998). Isoflavones, however, were not effective in lowering cholesterol in two other studies (Hodgson et al., 1998; Nestle et al., 1997). The exact mechanism by which soy exerts its hypocholesterolemic effect has not been fully elucidated.

Several classes of anticarcinogens have been identified in soybeans, including protease inhibitors, phytosterols, saponins, phenolic acids, phytic acid, and isoflavones (Messina and Barnes, 1991). Of these, isoflavones (genistein and daidzein) are particularly noteworthy because soybeans are the only significant dietary source of these compounds. Isoflavones are heterocyclic phenols structurally similar to the estrogenic steroids. Because they are weak estrogens, isoflavones may act as antiestrogens by competing with the more potent, naturally-occurring endogenous estrogens (e.g., 17b-estradiol) for binding to the estrogen receptor. This may explain why populations that consume significant amounts of soy (e.g., Southeast Asia) have reduced risk of estrogen-dependent cancer.

Soy may also benefit bone health. A clinical study involving 66 post-menopausal women conducted at the University of Illinois (Erdman and Potter, 1997) found that 40g isolated soy protein (ISP) per day (containing 90 mg total isoflavones) significantly increased (approximately 2%) both bone mineral content and density in the lumbar spine after 6 months. The theory that soy may alleviate menopausal symptoms was prompted by the observation that Asian women report significantly lower levels of hot flushes and night sweats compared to Western women. In a different study, 60 grams of ISP daily for 3 months reduced hot flashes by 45% in 104 postmenopausal women (Albertazzi et al., 1998). Although these observations are stimulating, there is a significant placebo effect in these studies, and it is too premature to suggest that soy may substitute for hormone replacement therapy.

Flaxseed: Among the major seed oils, flaxseed oil contains the most (57%) of the omega-3 fatty acid, a-linolenic acid. Recent research, however, has focused more specifically on fibre-associated compounds known as lignans. The two primary mammalian lignans, enterodiol and its oxidation product, enterolactone, are formed in the intestinal tract by bacterial action on plant lignan precursors (Setchell et al., 1981).

Because enterodiol and enterolactone are structurally similar to both naturally-occurring and synthetic estrogens, and have been shown to possess weakly estrogenic and antiestrogenic activities, they may play a role in the prevention of estrogen-dependent cancers. However, there are no epidemiological data and relatively few animal studies to support this hypothesis.

Fewer studies have evaluated the effects of flaxseed feeding on risk markers for cancer in humans. Adlercreutz et al. (1982) found that the urinary lignan excretion was significantly lower in postmenopausal breast cancer patients compared to controls eating a normal mixed or a lactovegetarian diet. Consumption of flaxseed has also been shown to reduce total and LDL cholesterol (Bierenbaum et al., 1993; Cunnane et al., 1993), as well as platelet aggregation (Allman et al., 1995).

Fruits

Acai (Euterpe oleracea Mart.) berry and other berries: The fruit of Euterpe oleraceae, commonly known as acai, has been demonstrated to exhibit significantly high antioxidant capacity in vitro, especially for superoxide and peroxyl scavenging, and, therefore, may have possible health benefits. A small, blackpurple palm fruit, it has been shown to reduce fasting glucose, insulin levels, total cholesterol and LDL-cholesterol, but not blood pressure, hs-CRP or eNO (endothelial nitric oxide) in healthy overweight adults (Udani, Singh, Singh, & Barrett, 2011). Anthocyanin (cyanidin 3-glucoside) compounds found in acai and other berries, may reduce blood glucose and enhanced insulin sensitivity by downregulating retinol binding protein expression in type 2 diabetic mice (Sasaki et al., 2007). Pure blueberries andstrawberry anthocyanins (but not whole berries, purplecorn anthocyanins or black raspberry anthocyanins) preventeddyslipidemia and obesity development in mammalson high fat diet (Prior, Wilkes, Rogers, Khanal, Wu, &Hager, 2010).

Berries are rich sources of a wide variety of antioxidant phenolics; these phytochemicals include flavonoids, stilbenes, tannins, and phenolic acids. Reactive oxidant species and free radicals are produced in an extensive range of physiological processes. In addition to the antioxidant defenses produced in the body, there are exogenous sources supplied by the diet; this is the case of berry fruits, among others. The insufficiency of antioxidant defense mechanisms is associated to the pathology of chronic disorders such as cardiovascular diseases, inflammation, and diabetes. Therefore, the enforcement of the latter mechanisms is of the utmost importance. The isolation and characterization of compounds that may delay the onset of aging is receiving intense research attention; some berry phenolics are being associated with this functional performance. Berry phenolics may also act as antimicrobials which may be of help in the control of the wild spectra of pathogens, in view of recent problems associated with antibiotic resistance. Most of the research works on the antioxidant activity of bioactive constituents of berries has been carried out using in vitro assays. In view of this, the human studies investigating the bioavailability and potential toxicity of phenolics are receiving more attention, although promising health benefits have been identified as describes in the paragraphs below.

Bilberries and to a lesser extent seabuckthorn decreased waist circumference, weight, vascular cell adhesion molecule and intercellular adhesion molecule of overweight and obese women (Lehtonen et al., 2011). Bilberry anthocyanidins-enriched extracts inhibited adipocyte differentiation by affecting the genes expressions of the insulin pathway; decreased peroxisome proliferator-activated receptor (PPAR) and sterol regulatory element-binding protein 1c (SREBP1c) and tyrosine residues of IRS1 phosphorylation (Suzuki et al., 2011). Blueberries (Vaccinium) help prevent degenerative diseases like diabetes, hyperlipidemia, hypertension, obesity, cancer, neurodegeneration and osteoporosis, through its apoptosis, antioxidant, antiinflammation and antiangiogenesis effects. Blueberries also improve visual function by increasing rhodopsin regeneration and have antimicrobial actions to prevent urinary tract infections. Blueberries bioactives include anthocyanins (anthocyanidins, or phenolic aglycone, conjugated with sugar), chlorogenic acid, flavonids, a-linolenic acid, pterostilbene, resveratrol and vitamins, that passes through the blood-brain barrier (Chen, Li, & Xu, 2010). Blueberries improved insulin sensitivity in obese, insulin-resistant men and women, without changing adiposity, energy intake and inflammatory biomarkers (Stull, Cash, Johnson, Champagne, & Cefalu, 2010). Blueberry attenuated insulin resistance in mammals by reducing adipocyte death, inflammation and insulin resistance (DeFuria et al., 2009). Blueberry reduced insulin resistance, triglycerides, fasting insulin, abdominal fat mass, liver weight, body weight and total fat mass, but increased adipose and skeletal muscle

PPAR activity and affected PPAR transcripts involved in fat oxidation and glucose uptake/oxidation in obese rats (Seymour et al., 2011). Blueberry juice and especially purified blueberry anthocyanins retarded obesity, serum leptin elevation and diabetes development in mice on obesogenic diet (Prior, Wilkes, Rogers,Khanal, Wu, & Howard, 2010). Blueberry polyphenols suppressedadipocyte differentiation, adipogenesis and cellproliferation (Moghe, Juma, Imrhan, & Vijayagopal, 2012). Blueberry anthocyanins reduced glucose production in hepatocytes, but did not increase glucose uptake (Roopchand, Kuhn, Rojo, Lila, & Raskin, 2013). Chokeberry reduced weight gain and modulates insulin, adipogenic and inflammatory signaling pathways in epididymal adipose tissue of rats on a fructose-rich diet (Qin & Anderson, 2012). Citrus polyphenols (from red orange, grapefruit, orange) produced lipolytic effect (via cAMP-phosphodiesterase inhibition) in overweight human adipocytes, more potently than cyanidin-3 glycoside, narangin or caffeine (Dallas, Gerbi, Tenca, Juchaux, &Bernard, 2008). Cranberry flavonoids ameliorated (i) insulin resistance (ii) plasma lipid profile, (iii) visceral fat mass, via the adiponectin-AMPK pathway, and (iv) plasma atherogenic cholesterol by downregulating the hepatic cholesterol synthesis pathway (Shabrova et al., 2011).

Fruit and vegetable consumption is inversely related to the incidence of heart disease and several cancers. However, many people in countries in Northern latitudes do not eat the recommended "5-a-day" of fruit and vegetables. For such populations, a potentially important source of fruit may be locally grown soft fruits (eg. raspberries, blackberries, blueberries, blackcurrants). Such berries contain micronutrients such as vitamin C and folic acid which are essential for health. However, berries may have additional health benefits as they are also rich in phytochemicals such as anthocyanins which are glycosidic-linked flavonoids responsible for their red, violet, purple and blue colours. In vitro studies indicate that anthocyanins and other polyphenols in berries have a range of potential anti-cancer and heart disease properties including antioxidant, anti-inflammatory, and cell regulatory effects. Such experimental data has lead to numerous health claims implying that "berries are edible superstars that may protect against heart disease, cancers and ageing." However, the bioavailabilty of polyphenols such as anthocyanins would appear to be limited, thus compromising their nutritional relevance

Apples: Apples are a widely consumed, rich source of phytochemicals, and epidemiological studies have linked the consumption of apples with reduced risk of some cancers, cardiovascular disease, asthma, and diabetes. In the laboratory, apples have been found to have very strong antioxidant activity, inhibit cancer cell proliferation, decrease lipid oxidation, and lower cholesterol. Apples contain a variety of phytochemicals, including quercetin, catechin, phloridzin and chlorogenic acid, all of which are strong antioxidants. The phytochemical composition of apples varies greatly between different varieties of apples, and there are also small changes in phytochemicals during the maturation and ripening of the fruit. Storage has little to no effect on apple phytochemicals, but processing can greatly affect apple phytochemicals. Apple consumption improved the lipid and oxidative status of the tissues and organs by producing larger intestinal pool and greater fecal excretion of bile acids in obese rats (Aprikian et al., 2002). Dietary apples also reduced the retroperitoneal and epididymal adipose tissue weights in rats (Nakazato, Song, & Waga, 2006).

Apples are a very significant source of flavonoids in people's diet in the US and in Europe. In the United States, twenty-two percent of the phenolics consumed from fruits are from apples making them the largest source of phenolics. In Finland, apples and onions are main sources of dietary flavonoids, while in the Netherlands apples rank third behind tea and onions as top sources of flavonoids. In a Finnish study of approximately 10,000 people, flavonoid intake was associated with a lower total mortality. Apples were one of the main sources of dietary flavonoids that showed the strongest associations with decreased mortality.

Not only are apples commonly enjoyed by many cultures, but they are also a good source of antioxidants. When compared to many other commonly consumed fruits in the United States, apples had the second highest level of antioxidant activity. Apples also ranked the second for total concentration of phenolic compounds, and perhaps more importantly, apples had the highest portion of free phenolics when compared to other fruits. This means that these compounds are not bound to other compounds in the fruits, and the phenolics may be more available for eventual absorption into the bloodstream.

Since fruits and vegetables are high in antioxidants, a diet high in these foods should help prevent oxidative stress, and may therefore help prevent chronic disease and slow aging. These findings have lead to the National Research Council of the United States to recommend consuming five or more servings of fruits and vegetables a day. Several commonly consumed foods and beverages, including tea, wine, onions, cocoa, cranberries, and apples, have been targeted as particularly beneficial in the diet because of their high content of phenolic compounds. While active research into the health benefits of these foods has been ongoing, current reviews of this work exist for all of the above-mentioned foods except apples.

Based on these epidemiological studies, it appears that apples may play a large role in reducing the risk of a wide variety of chronic disease and maintaining a healthy lifestyle in general. Apples are consistently associated with reduced risk of cancer, heart disease, asthma, and type II diabetes when compared to other fruits and vegetables and other sources of flavonoids. Apple consumption is also positively associated with increased lung function and increased weight loss. Partially because of such strong epidemiological evidence supporting the health benefits in apples, there is increasing research using animal and in vitro models that attempts to more clearly explain these health benefits.

Tomatoes: Selected by Eating Well magazine as the 1997 Vegetable of the Year, tomatoes have received significant attention within the last years because of interest in lycopene, the primary carotenoid found in this fruit (Gerster, 1997), and its role in cancer risk reduction.Lycopene gives tomatoes, pink grapefruit, watermelon, and guava their red color. It has been estimated that 80% of the lycopene in the US diet comes from tomatoes and tomato products like tomato sauce, tomato paste, and ketchup (Clinton, 1998). Lycopene is not a provitamin A carotenoid because it cannot be converted to retinol. Some foods that are good sources of lycopene are listed in Table 11 (USDA, 2015).

In a prospective cohort study of more than 47,000 men, those who consumed tomato products 10 or more times per week had less than one-half the risk of developing advanced prostate cancer (Giovannucci et al., 1995). Other cancers whose risk have been inversely associated with serum or tissue levels of lycopene include breast, digestive tract, cervix, bladder, and skin (Clinton, 1998). Proposed mechanisms by which lycopene could influence cancer risk are related to its antioxidant function. Lycopene is the most efficient quencher of singlet oxygen in biological systems (Di Mascio et al., 1989). The antioxidant function of lycopene may also explain the observation in a multi-center European study that adipose tissue levels of carotenoids were inversely associated with risk for myocardial infarction (Kohlmeier et al., 1997).

In 2004, FDA received two petitions for qualified health claims regarding tomatoes and/or lycopene and cancer risk reduction from The Lycopene Health Claim Coalition (consisting of H. J. Heinz Company, LycoRed Natural Products Industries, Ltd, The Morningstar Company, and The Prostate Cancer Foundation) and American Longevity, Inc. Both petitioners requested that FDA evaluate the relationship between tomato and/or lycopene consumption and prostate cancer risk. One petitioner also requested that FDA review the relationship between tomato and/or lycopene consumption and the risks of other forms of cancer, including lung, colorectal, gastric, breast, cervical, ovarian, endometrial, and pancreatic cancers. In response to these petitions, FDA evaluated evidence for associations between lycopene (a food component) and tomatoes (a food) separately and each form of cancer.

Based on FDA's consideration of the scientific evidence discussed above, the agency concluded that there was no credible evidence to support qualified health claims for tomatoes or tomato-based foods and a reduced risk for lung, colorectal, breast, cervical, or endometrial cancer. FDA further concluded that there was no credible evidence to support qualified health claims for lycopene, as a food ingredient, component of food, or as a dietary supplement, and a reduced risk of any of these cancers. Thus, FDA denied these claims. FDA concluded that there was very limited credible evidence for qualified health claims for tomatoes and/or tomato sauce and a reduced risk for prostate, gastric, ovarian, and pancreatic cancers provided that the qualified health claims were appropriately worded so as to not mislead consumers.

Citrus fruit: Several epidemiological studies have shown that citrus fruits are protective against a variety of human cancers. Although oranges, lemons, limes, and grapefruits are a principal source of such important nutrients as vitamin C, folate, and fibre, Elegbede et al. (1993) have suggested that another component is responsible for the anticancer activity. Citrus is most commonly thought of as a good source of vitamin C. However, like most other whole foods, citrus fruits also contain an impressive list of other essential nutrients, including both glycaemic and non-glycaemic carbohydrate (sugars and fibre), potassium, folate, calcium, thiamin, niacin, vitamin B6, phosphorus, magnesium, copper, riboflavin, pantothenic acid and a variety of phytochemicals. In addition, citrus contains no fat or sodium and, being a plant food, no cholesterol. The average energy value of fresh citrus is also low. For example a medium orange contains 60 to 80 kcal, a grapefruit 90 kcal and a tablespoon (15 ml) of lemon juice only 4 kcal (Whitney and Rolfes, 1999).

Citrus fruits are particularly high in a class of phytochemicals known as the limonoids (Hasegawa and Miyake, 1996). Evidence has been accumulating in support of the cancer preventative effect of limonene (Gould, 1997). Based on these observations, and because it has little or no toxicity in humans, limonene has been suggested as a good candidate for human clinical chemoprevention trial evaluation

Cranberry: Cranberry juice has been recognized as efficacious in the treatment of urinary tract infections since 1914, when Blatherwick (1914) reported that this benzoic acid-rich fruit caused acidification of the urine. Many investigations have focused on the ability of cranberry juice to inhibit the adherence of Escherichia coli to uroepithelial cells (Schmidt and Sobota, 1988). This phenomenon has been attributed to two compounds: fructose and a nondialyzable polymeric compound. The latter compound, subsequently isolated from cranberry and blueberry juices (Ofek et al., 1991), was found to inhibit adhesins present on the pili of the surface of certain pathogenic E. coli.

Avorn et al. (1994) published the results of the first randomized, double-blind, placebo-controlled clinical trial designed to determine the effect of a commercial cranberry juice beverage on urinary tract infections. One hundred fifty-three elderly women consuming 300 mL of cranberry beverage per day had significantly reduced (58%) incidence of bacteriuria with pyuria compared to the control group after six months. Based on the results of these studies, prevailing beliefs about the benefits of cranberry juice on the urinary tract appear to be justified.

Wine and Grapes

There is growing evidence that wine, particularly red wine, can reduce the risk of CVD. The link between wine intake and CVD first became apparent in 1979 when St. Leger et al. (1979) found a strong negative correlation between wine intake and death from ischemic heart disease in both men and women from 18 countries. France in particular has a relatively low rate of CVD despite diets high in dairy fat (Renaud and de Lorgeril, 1992). Although this "French Paradox" can be partly explained by the ability of alcohol to in-crease HDL cholesterol, more recent investigations have focused on the non-alcohol components of wine, in particular, the flavonoids.

The high phenolic content of red wine, which is about 20-50 times higher than white wine, is due to the incorporation of the grape skins into the fermenting grape juice during production. Kanner et al. (1994) showed that the black seedless grapes in red wines (i.e., Cabernet Sauvignon and Petite Sirah) contain high concentrations of phenols: 920, 1800, and 3200 mg/L, respectively, while green Thomson grapes contain only 260 mg/kg. Frankel and coworkers (1993) attributed the positive benefits of red wine to the ability of phenolic substances to prevent the oxidation of LDL, a critical event in the process of atherogenesis.

Although the benefits of wine consumption on CVD risk reduction seem promising, a recent prospective study of 128,934 adults in Northern California concluded that the benefits of alcohol consumption on coronary risk were not especially associated with red wine (Klatsky et al., 1997). Moreover, a note of caution is in order, as alcoholic beverages of all kinds have been linked to increased risk of several types of cancer, including breast cancer (Bowlin et al., 1997). Moderate wine consumption has also been associated with a decreased risk of age-related macular degeneration (Obisesan et al., 1998).

Those who desire health benefits of wine without potential risk may wish to consider alcohol-free wine, which has been shown to increase total plasma antioxidant capacity (Serafini et al., 1998). Furthermore, Day et al. (1998) showed that commercial grape juice is effective in inhibiting the oxidation of LDL isolated from human subjects. Red wine is also a significant source of trans-resveratrol, a phytoalexin found in grape skins (Creasy and Coffee, 1988). Resveratrol has also been shown to have estrogenic properties (Gehm et al., 1997) which may explain in part the cardiovascular benefits of wine drinking, and it has been shown to inhibit carcinogenesis in vivo (Jang et al., 1997).

Grapes or grape products contain numerous polyphenols, including the stilbene resveratrol, the flavanol quercetin, catechins and anthocyanins that help mitigate diabetes, by improving B-cell function and protecting against B-cell loss (Zunino, 2009). Grape-seed procyanidins modulated inflammation on human differentiated adipocytes by inhibiting NF-KB translocation to the nucleus; and reduces IL-6 and MCP-1 expression and enhances the anti-inflammatory adipokine adiponectin production (Chacon et al., 2009). Resveratrol formed by injured grapes, produced these benefits (Rosenow et al., 2012; Xu & Si, 2012)::

- calorie restriction effects
- improved human adipocyte secretion profile in obesity-induced metabolic disorders
- elevated basal glycerol release
- reduced intracellular TG content
- increased intracellular lipolysis
- down-regulated extracellular matrix proteins
- up-regulated processing proteins
- induced protective proteins secretion against cellular stress and apoptosis regulating proteins
- upregulated adiponectin and ApoE
- down-regulated PAI-1 and PEDF secretion which may improve anti-inflammatory processes
- increased insulin sensitivity
- inhibited cyclic adenosine monophosphate-specific phosphodiesterases
- activated 5'-adenosine monophosphate-activated kinase

Vitisin a resveratrol tetramer, strongly and dose dependently reduced (i) adipocyte differentiation and fat accumulation by cell cycle arrest through p21, (ii) PPARg expression (iii) Rb phosphorylation level and (iv) cell cycle at the G1 S phase transition, causing cells to remain in the preadipocyte state (Kim et al., 2008).

Vegetables

Onions and garlic family (Allium): Onion (Allium cepa L.) and garlic (Allium sativum L.), contain pungent thiosulfinates, phenolics, steroidal and volatile sulfur compounds, that may hinder cancer, coronary heart disease, obesity, hypercholesterol, type 2 diabetes, hypertension, cataract and gastrointestinal tract ailments (e.g. colic pain, flatulent colic and dyspepsia) (Lanzotti, 2006). Both Allium species contains N-acetylcysteine, which improved the antioxidant defences (superoxide dismutase, catalase, glutathione peroxidase, reduced glutathione) and mitigated glucose intolerance, dyslipidemia, LDLoxidation and serum oxidative stress (Souza et al., 2011). Onion dose-dependently reduced (i) diabetes related liver damage, (ii) blood glucose levels and (iii) lipid peroxidation product malondialdehyde (MDA), in diabetes animal models (Ogunmodede, Saalu, Ogunlade, Akunna, & Oyewopo, 2012). Onion has hypoglycemic, cardioprotective, hypolipidemic and bone strengthening properties. Onion is rich in flavonoids, polyphenols, organic sulfur and saponins. Onion is more potent in its raw form (Sohail, Karim, Sarwar, & Alhasin, 2011). Garlic works synergistically with type 2 antidiabetic agents (Ashraf, Khan, & Ashraf, 2011). Garlic and propranolol are potent antihypertensive and cardioprotective agents and when combined, garlic doubled or tripled the bioavailability and half life of propranolol. Garlic effectively reduced fluid intake, body weight, blood pressure, cholesterol, triglycerides and blood glucose (Asdaq & Inamdar, 2011). Garlic acts synergistically with turmeric against hyperglycemia, HbA1C and dyslipidemia in type-2 diabetesdyslipidemia patients (Sukandar et al., 2010).

Garlic (Allium sativum) is likely the herb most widely quoted in the literature for medicinal properties (Nagourney, 1998). The claimed health benefits of garlic are numerous, including its role as cancer chemopreventive, antibiotic, anti-hypertensive, and cholesterol-lowering (Srivastava et al., 1995).

The characteristic flavour and pungency of garlic are due to an abundance of oil-and water-soluble, sulfur-containing elements, which are also likely responsible for the various medicinal effects ascribed to this plant. However, intact, undisturbed bulbs of garlic contain only a few medicinally active components. The intact garlic bulb contains an odourless amino acid, alliin, which is converted enzymatically by allinase into allicin when the garlic cloves are crushed (Block, 1992b). This latter compound is responsible for the characteristic odor of fresh garlic. Allicin then spontaneously decomposes to form numerous sulfur-containing compounds, some of which have been investigated for their chemopreventive activity.

Garlic components have been shown to inhibit tumorigenesis in several experimental models (Reuter et al., 1996). However, additional reports have shown garlic to be ineffective. Inconclusive results are likely due to differences in the type of garlic compounds or preparations used by various investigators. Considerable variation in the quantity of organosulfur compounds available in fresh and commercially available garlic products has been demonstrated (Lawson et al., 1991).

Several epidemiologic studies show that garlic may be effective in reducing human cancer risk (Dorant et al., 1993). A relatively large case-control investigation conducted in China showed a strong inverse relationship between stomach cancer risk and increasing allium intake (You et al., 1988). In a study of more than 40,000 postmenopausal women, garlic consumption was associated with nearly a 50% reduction in colon cancer risk (Steinmetz et al., 1994).

Not all epidemiological studies, however, have shown garlic to be protective against carcinogenesis. A 1991 review of 12 case-control studies (Steinmetz and Potter, 1991b), found that eight showed a negative association, one showed no association, and three studies showed a positive association. A review of 20 epidemiological studies (Ernst, 1997) suggests that allium vegetables, including onions, may have a protective effect on cancers of the gastrointestinal tract.

Garlic has also been advocated for the prevention of CVD, possibly through antihypertensive properties. According to Silagy and Neil (1994), however, there is still insufficient evidence to recommend it as a routine clinical therapy for the treatment of hypertensive subjects. The cardioprotective effects are more likely due to its cholesterol-lowering effect. In a meta-analysis, Warshafsky et al. (1993) summarized the results of five randomized, placebo-controlled clinical trials, involving 410 patients. They showed that an average of 900 mg garlic/day (as little as one half to one clove of garlic) could decrease total serum cholesterol levels by approximately 9%. The validity of both of these reports, however, is reduced by methodological shortcomings, including the fact that dietary intake, weight, and/or exogenous garlic ingestion was not always well-controlled. In a multi-center, randomized, placebo-controlled trial in which dietary assessment and supervision were strictly controlled, 12 weeks of garlic treatment was ineffective in lowering cholesterol levels in subjects with hypercholesterolemia (Isaacsohn et al.,

1998). It is currently unclear which component in garlic is responsible for its cholesterol-lowering effect.

Brassica family: Brassicas (cabbage, cauliflower, broccoli, broccolini,brussels sprouts, chinese cabbage, swedes and turnips,broccoli raab, collards, cress, kale, kohlrabi, mustard, kalianand bok choi) contain glucosinolates, carotenoids,chlorophylls, ascorbic acid, sinigrin; gluconapin, glucobrassicanapinand progoitrin, phenolics like hydroxycinnamicacids (sinapic, ferulic, p-coumaric and caffeicacids), glycosides of quercetin and kaempferol and derivativesof p-coumaric, ferulic, sinapic and caffeic acid(Krumbein, Schonhof, & Schreiner, 2005; Olsen, Aaby, &Borge, 2009).

Epidemiological evidence has associated the frequent consumption of cruciferous vegetables with decreased cancer risk. In a review of 87 case-control studies, Verhoeven et al. (1996) demonstrated an inverse association between consumption of total brassica vegetables and cancer risk. The percentages of case-control studies showing an inverse association between consumption of cabbage, broccoli, cauliflower, and Brussels sprouts and cancer risk were 70, 56, 67, and 29%, respectively. Verhoeven et al. (1996) attributed the anticarcinogenic properties of cruciferous vegetables to their relatively high content of glucosinolates.

Glucosinolates are a group of glycosides stored within cell vacuoles of all cruciferous vegetables. Myrosinase, an enzyme found in plant cells, catalyses these compounds to a variety of hydrolysis products, including isothiocyanates and indoles. Indole-3 carbinol (I3C) is currently under investigation for its cancer chemopreventive properties, particularly of the mammary gland. In addition to the induction of phase I and II detoxification reactions, I3C may reduce cancer risk by modulating estrogen metabolism. The C-16 and C-2 hydroxylations of estrogens involve competing cytochrome P-450-dependent pathways, each sharing a common estrogen substrate pool. Studies suggest that the increased formation of 2-hydroxylated (catechol) estrogen metabolites relative to 16-hydroxylated forms, may protect against cancer, as catechol estrogens can act as antiestrogens in cell culture. In contrast, 16-hydroxyestrone is estrogenic and can bind to the estrogen receptor. In humans, I3C administered at 500 mg daily (equivalent to 350-500 g cabbage/day) for 1 week significantly increased the extent of estradiol 2-hydroxylation in women (Michnovicz and Bradlow, 1991), suggesting that this compound may be a novel approach for reducing the risk of breast cancer.

Although a wide variety of naturally occurring and synthetic isothiocyanates have been shown to prevent cancer in animals (Hecht, 1995), attention has been focused on a particular isothiocyanate isolated from broccoli, known as sulforaphane. Sulforaphane has been shown to be the principal inducer of a particular type of Phase II enzyme, quinone reductase. Fahey et al., (1997) recently demonstrated that 3-day-old broccoli sprouts contained 10-100 times higher levels of glucoraphanin (the glucosinolate of sulforaphane) than did corresponding mature plants. However, in view of the importance of an overall dietary pattern in cancer risk reduction, the clinical implications of a single phytochemical in isolation has been questioned (Nestle, 1998).

*Beetroot (Beta vulgaris var. rubra):*In recent years, there has been a growing interest in the biological activity of red beetroot (Beta vulgaris rubra) and its potential utility as a health promoting and disease preventing functional food. As a source of nitrate, beetroot ingestion provides a natural means of increasing in vivo nitric oxide (NO) availability and has emerged as a potential strategy to prevent and manage pathologies associated with diminished NO bioavailability, notably hypertension and endothelial function. Beetroot is also being considered as a promising therapeutic treatment in a range of clinical pathologies associated with oxidative stress and inflammation. Its constituents, most notably the betalain pigments, display potent antioxidant, anti-inflammatory and chemo-preventive activity in vitro and in vivo. Beetroot has oxidative, proapoptotic and anti-inflammatory effects in obese women neutrophils (Zielinska-Przyjemska, Olejnik, DobrowolskaZachwieja, & Grajek, 2009). Recent studies have provided compelling evidence that beetroot ingestion offers beneficial physiological effects that may translate to improved clinical outcomes for several pathologies, such as; hypertension, atherosclerosis, type 2 diabetes and dementia. Hypertension in particular has been the target of many therapeutic interventions and there are numerous studies that show beetroot as a fast growing functional food ingredient, delivered acutely as a juice supplement or in bread, significantly reduce systolic and diastolic blood pressure.

*Seaweed:*Seaweed as a potential functional food goes well beyond its role as a sush wrap. Edible seaweeds are rich in bioactive antioxidants, soluble dietary fibers, proteins, minerals, vitamins, phytochemicals, and polyunsaturated fatty acids. Although previously the seaweeds were only used as gelling and thickening agents in the food or pharmaceutical industries, recent researches have revealed their potential as complementary medicine. The red, brown and green seaweeds have been shown to have therapeutic properties for health and disease management, such as anticancer, antiobesity, antidiabetic, antihypertensive, antihyperlipidemic, antioxidant, anticoagulant, anti-inflammatory, immunomodulatory, antiestrogenic, thyroid stimulating, neuroprotective, antiviral, antifungal, antibacterial and tissue healing properties in vivo. Active compounds include sulphated polysaccharides, phlorotannins, carotenoids (e.g. fucoxanthin), minerals, peptides and sulfolipids, with proven benefits against degenerative metabolic diseases. Their therapeutic modes of action and bioactive components have been meticulously studied by scientists. Seaweed biomass is already used for a wide range of food products, including stabilising agents. Biorefineries with seaweed as feedstock are attracting worldwide interest and include low-volume, high value-added products and vice versa. Scientific research on bioactive compounds in seaweed usually takes place on just a few species and compounds. Seaweeds antioxidative fucoidans deter inflammation, allergy, tumor, obesity, coagulation, virus, hepatopathy, uropathy and renalpathy (Vo & Kim, 2013). Fucoidan suppresses lipid accumulation and adipocyte differentiation by suppressing inflammation-related cytokines, PPARg, CCAAR/enhancer-binding protein a, and adipocyte protein 2 (Kim & Lee, 2012).

Herbs and spices:

The pungent compounds in herbs and spices help enhance metabolism and deter against obesity, diabetes, and chronic inflammation. Many herbs and spices have good positive effects on blood sugar, insulin sensitivity, dyslipidaemia, weight gain and the cardiovascular system. They activate PPARa and PPARg, inhibit NF-KB activation, and enhance anti-inflammatory cytokines expression (Jungbauer & Medjakovic, 2012). Black cumin (Nigella sativa) reduced blood total cholesterol and weight gain in animal models (Tauseef, Butt, & Anjum, 2009). Catharanthus roseus, Thymus vulgaris, Hypericum perforatum, and Artemisia annua are also potently antioxidative. Rosmarinic acid was the main phenolic in Salvia officinalis, Thymus vulgaris, Origanum majoricum, and P. longiflora (Patel, Prasad, Kumar, & Hemalatha, 2011).

Cinnamon: Cinnamon extract helped safeguard against tissue changes by minimizing protein aggregation, preventing glycation and oxidative stress in fructose-fed rodents. Cumin, cinnamon, and black pepper, inhibited in vitro advanced glycation end-products (AGE) formation at 1.0 mg/ml (Saraswat et al., 2010). Piperine from black pepper (Piper nigrum) inhibited lipid peroxidation (Chonpathompikunlert, Wattanathorn, & Muchimapura, 2010). Cinnamon water extract or polyphenols with doubly linked procyanidin type-a polymers display insulin-like activity to improve blood glucose and lipid profiles of diabetes patients. Cinnamon increased proteins for insulin signaling, glucose transport and anti-inflammatory/antiangiogenesis response (Cao, Polansky, & Anderson, 2007). Metabolic syndrome subjects consuming 1-6g cinnamon had improved fasting blood glucose, systolic blood pressure, body fat and lean body mass compared with the placebo group (Anderson, 2008). Cinnamon normalizes fasting blood glucose by decreasing hepatic glucose production and the gene expression of related enzymes (phosphoenolpyruvate carboxykinase and glucose-6-phosphatase) in obese hyperglycemic mice (Cheng et al., 2012).

Clove: Clove aqueous extracts have insulin-like biological activity (Broadhurst, Polansky, & Anderson, 2000) and 120 S. Mohamed / Trends in Food Science & Technology 35 (2014) can gradually lower fasting blood glucose levels (Shukri, Mohamed, & Mustapha, 2010). The eugenol and eugenyl acetate in cloves are anti-oxidative. Fenugreek (Trigonella foenum-graecum Linn) seeds alcoholic extract has beneficial effects on body weight, blood glucose and cataract development in aging humans.

Ginger: Ginger (Zingiber officinale) is a perennial plant with narrow, bright green, grass-like leaves and yellowish green flowers with purple markings. Ginger is cultivated in the tropics for its edible rhizome at approximately 10 months of age, with the root stocks serving a variety of purposes, including culinary and medicinal (Grant, 2000; Ursell, 2000; Portnoi et al. 2003). The efficacy of ginger is purported to be a result of its aromatic, carminative and absorbent properties (Govindarajan, 1982a,b). Ginger is a widely used spice and functional food. For centuries ginger has been an important ingredient in Chinese, Ayurvedic and Tibb-Unani herbal medicine. The medicinal properties attributed to ginger include anti-arthritic (Srivastava & Mustafa, 1989, 1992; Bliddal et al. 2000), anti-migraine (Mustafa & Srivastava, 1990; Cady et al. 2005), anti-thrombotic (Bordia et al. 1997; Thomson et al. 2002), anti-inflammatory (Thomson et al. 2002, Penna et al. 2003), hypolipidaemic (Bordia et al. 1997; Thomson et al. 2002; Bhandari et al. 2005), hypocholesterolaemic (Furhman et al. 2000; Bhandari et al. 2005) and anti-nausea properties (Ernst & Pittler, 2000; Portnoi et al. 2003).

Ginger is known to contain several potentially bioactive substances, mainly gingerols and their related dehydration products, the shogaols, as well as volatile oils including sesquiterpenes, such as b-bisabolene and (-)-zingiberene, and monoterpenes, mainly geranial and neral (Chevallier, 1996; Ursell, 2000; McKenna et al. 2002). In particular, gingerols have been shown to inhibit both prostaglandin and leukotriene biosynthesis (Kiuchi et al. 1992) and angiogenesis (Kim et al. 2005). In addition, several ginger components exhibit serotonin receptor-blocking activity (Huang et al. 1991; Abdel-Aziz et al. 2005). A few isolated studies about the hypoglycaemic properties of ginger in animals have reported variable results. Sharma & Shukla (1977) reported a small but significant blood glucose-lowering effect of ginger juice in diabetic and non-diabetic animals. Akhani et al. (2004) also observed that ginger juice exhibits hypoglycaemic activity in both normal and streptozotocin (STZ)-induced diabetic rats. Mascolo et al. (1989) reported a significant hypoglycaemic activity in normal rabbits at different times after a variety of administration schedules and doses. Ginger (Zingiber officinalis) is thermogenic, antioxidative, stimulating, anti-inflammatory, anti-hyperglycemic and inhibited glycation (Saraswat et al., 2010).

*Turmeric:*Turmeric cultivation in India occupies 60% of the total area intended for spices and condiments. It is not only an ancient but also a highly cultivated Asian spice and used in countries like India, China, Malaysia, Pakistan, Bangladesh, Indonesia, Taiwan, Haiti, Jamaica and El-Salvador. Traditionally turmeric is used in a variety ways for different diseases, due to its diverse properties, for example as a general tonic, stimulant, cosmetic and for the treatment of coughs, colds, soar throats, asthma and dyspepsia including peptic ulcers. It is also used as a deworming agent and as a paste for some viral diseases such as chicken pox, small pox and measles. It is liberally used in arthritis and for wound healing. Current literature abounds in scientific evidences supporting its traditional uses and evidence based science clearly shows that turmeric and its active constituents, particularly curcuminoids can be used for a variety of disorders.

Turmeric is a popular spice derived from the root of Curcuma longa Linn, a member of the ginger family. The powder and its active principle, a group of curcuminoids, are widely used as: culinary spices, preservatives, food additives, cosmetics, and as oleoresin in food and pharmaceutical industries. In the last two decades, there has been considerable interest among the biomedical scientists to explore the possible therapeutic benefits of turmeric and its active principle, curcuminoids. A considerable amount of work has been carried out to explain its molecular, cellular, biochemical and pathological mechanisms to establish its potential as a therapeutic agent for many chronic diseases. Turmeric contains the antioxidant curcumin (diferuloylmethane) that affect signaling mediators like NFKB, cyclooxygenase-2 (COX-2), lipooxygenase (LOX), and inducible nitric oxide synthase (iNOS), for modulating inflammation, angiogenesis, transcription, enzymes, protein kinases, protein reductases, carrier proteins, cell survival, resistance, adhesion, growth factors, receptors, cell cycle regulation, chemokines, DNA, RNA and metal ions. Turmeric helps prevent and/or treat arteriosclerosis, oxidation, respiratory, hepatic, pancreatic, and degenerative diseases in humans (Anand, Kunnumakkara, Newman, & Aggarwal, 2007). Turmeric and curcumin are not anti-hyperglycemic but prevented oxidative stress in hyperglycemic rats and induced the glutathione-linked detoxifi- cation enzymes in rats (Suryanarayana et al., 2005).

Curcumin: Curcumin at the 0.002% level inhibited oxidation, glycation, lipid peroxidation, AGE and protein aggregation. Under hyperglycemic conditions, higher levels of dietary curcumin (0.01%) may have the opposite effect of being pro-oxidative, enhancing AGE formation and protein aggregation. However feeding curcumin to normal rats at up to 0.01% level did not cause any changes in morphology or biochemical parameters (Suryanarayana et al., 2005) because of its limited bioavailability. Curcumin protective effects against lipid peroxidation or galactose induced oxidative stress are (i) through glutathione S-transferase isozyme induction and (ii) by decreasing cells apoptosis (Pandya et al., 2000). Dietary curcumin improved tissue morphology, physiology, and gene expression, tissue structure and function in animal models. Vitex negundo help normalize blood sugar and decreasing oxidative stress (Rooban et al., 2009). Other herbs like Dendrobium moniliforme is antioxidative, and helped normalize serum glucose, total cholesterol, renal lipid and renoprotective in high fat diet mice. The Dendrobium decreased high-fat diet-induced renal damage in mice through the regulation of lipid-induced oxidative stress (Lee et al., 2012). Cleistocalyx operculatus flower buds inhibit pancreatic lipase and a-amylase to help prevent or treat obesity (Zhang & Lu, 2012). Bergenia crassifolia leaves extract is an appetite suppressant that helps prevent obesity, and improve glucose tolerance without affecting cholesterol level (Shikov et al., 2012).

Table 11: Lycopene content of selected foods

Food	Serving	Lycopene (mg)
Tomato paste, canned	1 cup	75.4
Tomato purée, canned	1 cup	54.4
Tomato soup, canned, condensed	1 cup	26.4
Vegetable juice cocktail, canned	1 cup	23.3
Tomato juice, canned	1 cup	22.0
Watermelon, raw	1 wedge (1/16 of a melon that is 15 inches long x 7.5 inches in diameter)	13.0
Tomatoes, raw	1 cup	4.6
Ketchup (catsup)	1 tablespoon	2.5
Pink grapefruit, raw	½ grapefruit	1.7
Baked beans, canned	1 cup	1.3

Tea

Tea is second only to water as the most widely consumed beverage in the world. A great deal of attention has been directed to the polyphenolic constituents of tea, particularly green tea (Harbowy and Balentine, 1997). Green tea, a time-honoured Chinese herb, might be regarded as a functional food because of its inherent anti-oxidant, anti-inflammatory, antimicrobial and antimutagenic properties. They are attributed to its reservoir of polyphenols, particularly the catechin, epigallocatechin-3-gallate. Owing to these beneficial actions, this traditional beverage was used in the management of chronic systemic diseases including cancer. Recently, it has been emphasized that the host immuno-inflammatory reactions destroy the oral tissues to a greater extent than the microbial activity alone. Green tea with its wide spectrum of activities could be a healthy alternative for controlling these damaging reactions seen in oral diseases, specifically, chronic periodontitis, dental caries and oral cancer, which are a common occurrence in the elderly population. Polyphenols comprise up to 30% of the total dry weight of fresh tea leaves. Catechins are the predominant and most significant of all tea polyphenols (Graham, 1992). The four major green tea catechins are epigallocatechin-3-gallate, epigallocatechin, epicatechin-3-gallate, and epicatechin.

Tea is the most common functional beverage consumed worldwide. It is derived from the leaves of *Camellia sinensis*, which is processed in different ways to produce either green, black or oolong tea. Of the total tea produced, 20% is green tea, consumed commonly in Asia, parts of North America and the Middle East, whereas 78% is black tea popular in Western and Asian countries. Less than 2% is oolong tea, consumed mainly in southern China and Taiwan.

Green tea, technically known as *Camellia sinensis* Theaceae, is the richest source of polyphenols, specifically epigallocatechin-3-gallate (EGCG). It is a non-fermented form of tea produced by drying and steaming the fresh leaves to inactivate the polyphenol oxidase. Black tea contains thearubigins and theaflavins, which are produced from the enzymatic oxidation ("fermentation") of polyphenols in the crushed leaves of *Camellia sinensis*. Oolong tea is manufactured by partial oxidation of these leaves.

Studies have shown that green tea has better health effects over black tea. The polyphenol concentration in green tea is approximately 30–40% as compared with 3–10% in black tea, resulting in the highest anti-oxidant activity. Besides, it has potent anti-inflammatory, antibacterial, antiviral, antimutagenic and anti-aging properties.

There has been a great deal of interest in the pharmacological effects of tea. By far, most research on the health benefits of tea has focused on its cancer chemopreventive effects, although the epidemiological studies are inconclusive at the present time (Katiyar and Mukhtar, 1996). In a 1993 review of 100 epidemiological studies (Yang and Wang, 1993), approximately 2/3 of the studies found no relationship between tea consumption and cancer risk, while 20 found a positive relationship and only 14 studies found that tea consumption reduced cancer risk. A more recent review suggests that benefits from tea consumption are restricted to high intakes in high-risk populations (Kohlmeier et al., 1997b). This hypothesis supports the recent finding that the consumption of five or more cups of green tea per day was associated with decreased recurrence of stage I and II breast cancer in Japanese women (Nakachi et al., 1998).

In contrast to the inconclusive results from epidemiological studies, research findings in laboratory animals clearly support a cancer chemopreventive effect of tea components. In fact, Dreosti et al.(1997) stated that "no other agent tested for possible chemoprevention effects in animal models has elicited such strong activity as tea and its components at the concentrations usually consumed by humans."

There is some evidence that tea consumption may also reduce the risk of CVD. Hertog and coworkers (1993) reported that tea consumption was the major source of flavonoids in a population of elderly men in the Netherlands. Intake of five flavonoids (quercetin, kaempferol, myricetin, apigenin, and luteolin), most which was derived from tea consumption, was significantly inversely associated with mortality from CHD in this population. Although several other prospective studies have demonstrated a substantial reduction in CVD risk with tea consumption, the evidence is not presently conclusive (Tijburg et al., 1997).

Coffee

Coffee is a highly popular drink that is traditionally used to complement meals, as well as for hedonistic and psychostimulant purposes. It is not generally consumed for nourishment. However, depending on the quantities consumed, it can affect the intake of potassium, magnesium and manganese (Gillies & Birkbeck, 1983). Coffee can provide 8 % of the daily intake of chromium (Santos et al. 2004) and can be a substantial source of magnesium; a mean of 63.7mg/cup (100 ml) has been reported (Astier-Dumas & Gounelle de Pontanel, 1974). This is comparable to the upper range of magnesium concentrations (mean 28.8 mg/ml, range

1.7–95.4 mg/ml) reported for non-alcoholic drinks (JodralSegado et al. 2003). Indeed, magnesium excretion has been positively correlated with magnesium intake and coffee consumption (Lowik et al. 1993). The food service industry has introduced new, highenergy gourmet coffee beverages (Shields et al. 2004). In some circumstances, coffee-brewing can remove toxic metals such as lead from influent water (Impellitteri et al. 2000). The coffee beverage is rich in biologically active substances such as nicotinic acid, trigonelline, quinolinic acid, tannic acid, pyrogallic acid and, of course, caffeine (Minamisawa et al. 2004). Niacin in particular is formed in great amounts from trigonelline during the coffee-bean roasting process (Czok, 1977; Casal et al. 2000).

Coffee beans and tea both contain phenolic compounds and antioxidant compounds. However, due to the coffee bean-roasting process, phenolic compounds can be lost while compounds with antioxidant properties are being developed. Indeed, overall antioxidant properties may be enhanced by newly formed Maillard reaction products (MRP) during roasting (Nicoli et al. 1997, 1999; Anese & Nicoli, 2003). Nicoli et al. (1999) studied these products and demonstrated that beans that have been prepared using intermediate roasting conditions have maximum antioxidant activity. Decreases in antioxidant properties seem to be related to high-molecular weight MRP. However, during the storage of ready-to-drink brews, a further development of MRP occurs, but there is a decrease in antioxidant properties (Anese & Nicoli, 2003). Daglia et al. (2004) tested in vitro and ex vivo specific antiradical activity against hydroxyl radicals and found that 5-O-caffeoiquinic acid was the most active fraction. Dogasaki et al. (2002) reported that brewed coffee possessed antibacterial activities exhibited by caffeic acid, chlorogenic acid and protocatechnic acid (3,4-dihydoxy benzoic acid). Daglia et al. (1998, 2002) also reported anti-adhesive properties due to both naturally occurring and roasting-induced molecules.

Epidemiological and experimental studies have shown positive effects of regular coffee-drinking on various aspects of health, such as psychoactive responses (alertness, mood change), neurological (infant hyperactivity, Parkinson's disease) and metabolic disorders (diabetes, gallstones), and gonad and liver function. Unlike other functional foods that act on a defined population with a special effect, the wide use of coffee-drinking impacts a broad demographic (from children to the elderly), with a wide spectrum of health benefits. Many studies support the idea that coffee-drinking has health benefits. Thus, it is simple to conceptualise coffee as a functional food.

Coffee's most studied component, caffeine, varies substantially as a function of coffee plant species and the method of bean-roasting and drinks preparation. The caffeine content of caffeinated coffees ranges from 58 to 259 mg/dose. In one study, the mean caffeine content of brewed speciality coffees was 188 mg for a 16 oz (USA) cup (Bell et al. 1996). Variability is, however, high. McCusker et al. (2003) reported a wide range of caffeine concentration (259–564 mg/dose) in the same coffee beverage obtained from the same outlet on six consecutive days. The strong pharmacological effects of caffeine have led to consumer demand for caffeine-free coffee beverages. Many decaffeination techniques have been developed. At first, organic solvents were used, but this was discontinued due to their undesirable sideeffects. Decaffeination techniques employing supercritical CO_2 are currently used to extract caffeine from coffee beans. McCusker et al. (2003) analysed coffee brands sold as decaffeinated and found them to have caffeine concentrations less than 17.7 mg/ dose. The ability of newly developed coffee plants (naturally selected or GM) to possess a low caffeine content while keeping the rich flavour and aroma of their caffeinated counterparts holds promise. In fact, these transgenic plants showed a 70 % reduction in caffeine content (Ogita et al. 2003), whereas a naturally decaffeinated Coffea arabica plant (genetic selection) from Ethiopia showed a 93.7 % reduction in the level of caffeine (Silvarolla et al. 2004).

Animal sources of Functional Foods

Although the vast number of naturally occurring health-enhancing substances are of plant origin, there are a number of physiologically-active components in animal products that deserve attention for their potential role in optimal health.

Fish

Fish and other seafood are fundamentally functional and possess many beneficial components for human health. Seafood and its by-products are an excellent source of functional food componenets. These can be extracted and isolated and added to a range of foods, thereby enhancing functionality of the foods in terms of human health (Shahidi, 2003; Alasalvar and Taylor, 2002).

The market for seafood products worldwide has grown significantly in recent years largely fuelled by the image of fish as a healthy component of the diet (Kinsella, 1986; Anon., 2001). Fish and fish products are increasingly promoted as functional foods (Gormley, 2006), and Alasalvar and Taylor (2002) have reviewed applications of marine ingredients to food and health. Many health benefits have, and are, being attributed to fish and include risk reduction and management of a number of diseases and chronic conditions including cardiovascular health and blood pressure; blood clotting; cancer; arthritis; vitamin and mineral deficiencies (Gormley, 2006).

Omega-3 (n-3) fatty acids are an essential class of polyunsaturated fatty acids (PUFAs) derived primarily from fish oil. It has been suggested that the Western-type diet is currently deficient in n-3 fatty acids, which is reflected in the current estimated n-6 to n-3 dietary ratio of 20:25, compared to the 1:1 ratio on which humans evolved (Simopoulos, 1991). This has prompted researchers to examine the role of n-3 fatty acids in a number of diseases (particularly cancer and CVD) and more recently, in early human development.

That n-3 fatty acids may play an important role in CVD was first brought to light in the 1970s when Bang and Dyerberg (1972) reported that Eskimos had low rates of this disease despite consuming a diet which was high in fat. The cardioprotective effect of fish consumption has been observed in some prospective investigations (Krumhout et al., 1985), but not in others (Ascherio et al., 1995). Negative results could be explained by the fact that although n-3 fatty acids have been shown to lower triglycerides by 25-30%, they do not lower LDL cholesterol. In fact, a recent review of 72 placebo- controlled human trials, showed that n-3 fatty acids may increase LDL cholesterol (Harris, 1996).

Although eating large amounts of fish has not unequivocally been shown to reduce CVD risk in healthy men, consumption of 35 g or more of fish daily has been shown to reduce the risk of death from nonsudden myocardial infarction in the Chicago Western Electric Study (Daviglus et al., 1997), and as little as one serving of fish per week was associated with a significantly reduced risk of total cardiovascular mortality after 11 years in more than 20,000 U.S. male physicians (Albert et al., 1998).

There are increasing demands for healthy convenience fish products and this gives scope for enriching fish products with dietary fibres (DFs), antioxidant dietary fibres (AODFs) or with prebiotics, thereby making the products doubly functional, i.e. the inherent functionality of the fish and the added functionality of the DF, AODF or prebiotic. AODF is plant derived material with a DF content of at least 50% (dry matter basis), an inhibitory lipid oxidation potential (per g AODF) >200 mg of vitamin E, and free radical scavenging capacity >50 mg vitamin E. Examples of some products containing DF, AODF or a prebiotic are cited below.

- Apple pomace in mackerel products: Apple pomace is a cheap product with minimal commercial value and is produced in volume in many countries but especially in Poland where there is a large production of apples and a significant fruit juice industry. Pomace from the apple cultivar Shampion is white and, therefore, conducive to being adding to products. Pomace is largely cell wall material with significant pectin content, and is known to reduce cholesterol and low density LDL cholesterol in addition to its other properties as an AODF (Dragsted and Gormley, 2010). Keilen et al. (2007) added pomace (0, 2, 4% w/w) to minced mackerel tissue (Scomber scombrus) and this was formed into burgers. The pomace had a DF content of 50%; total phenols were 569 mg gallic acid equivalents /100 g dw, but the anti radical power of 0.312 was low. Sensory acceptability scores were >3 (mid-point of scale) with the exception of the sample with 4% apple pomace on day 7. Thiobarbituric acid substances (TBARS) values were high suggesting oxidation of the fish lipid but the analysis was influenced by the apple pomace itself and by other constituents in the burgers as indicated by increasing TBARS values with rising inclusions of apple pomace. In addition, TBARS values rose over time indicating increasing oxidation and that the pomace was not performing efficiently as an antioxidant under the conditions prevailing in the trial.
- Wheat, grape and seaweed DF inclusions in seafood products: Careche et al. (2008) incorporated wheat DF (WDF) into minced fish at levels up to 6% (w/w) and/or to surimi gels and non-gelled products (3-6% w/w inclusions). In many of the

trials 3% WDF was the upper level of inclusion in terms of product sensory acceptability. Grape and seaweed AODFs were also tested in-product. The former was included in restructured products made from horse mackerel (Sanchez-Alonso et al., 2007). Tests indicated that inclusions of 2% in-product were well accepted sensorically, but that 4% gave a less acceptable texture but did not impact on product flavour. Tests were also conducted on including seaweed AODF extracted from fucus vesiculosus into restructured seafood products. However, the inclusions imparted an undesirable green colour to the product and also an over-strong flavour (Careche et al., 2008).

- Fish ready-meals and fish-fruit combination products: While fish is an excellent and nutritious product,its use in combination with other food types also provides major commercial opportunities. For example, Braida and Gormley (2008) produced a gluten-free salmon lasagne ready-meal containing an oligosaccharide (a prebiotic and sweetener), taurine, pectin or algal calcium, or a combination of all four. Gluten-free is also a requirement for an increasing number of consumers. The meal comprised salmon sauce, (50% salmon pieces), Bėchamel sauce, gluten-free pasta sheets and mozzarella cheese and was based on conventional salmon lasagne. The functional food ingredients were included in the Bėchamel sauce at a range of concentrations. Each of the inclusions reduced sauce acceptability score for the Bėchamel sauce, and the all-inclusions combined treatment (pectin content 2.5%) got the lowest sensory score. The Bėchamel sauce with a prebiotic got a similar rating to the control on a 6-cm line scale with end-points 0 (unacceptable) and 6 (very acceptable) and these data transferred to the acceptability of the lasagne ready-meals. The conclusion was that the gluten-free lasagne ready-meals with prebiotic, taurine or pectin can be classed as new-generation products targeted at the health market.
- Conjugated linoleic acid inclusions in seafoods: Conjugated linoleic acid (CLA) has been linked to a reduction of body fat without substantially reducing body weight. In-vitro studies have shown that CLA may have anti-carcinogenic activity,

protect against atherosclerosis, and have other positive physiological effects (Pariza et al., 2001; Belury, 2002; Noone et al., 2002). There has, therefore, been extensive interest in the last five years in enriching fish fillets with CLA thereby giving additional functionality. Manning et al. (2006), enriched catfish (Ictalurus punctatus) with CLA and ω-3 PUFAs by dietary manipulation. Fish muscle showed an increase in both CLA and ω-3 PUFAs compared to controls and the content of CLA was 6.4 % of total lipid indicating successful enrichment. Valente et al. (2007a) studied the effect of CLA (diets contained up to 2%) on growth, lipid composition and hepatic lipogenesis in juvenile European sea bass (Dicentrarchus labrax). The total accumulation of CLA was 5.6% in the muscle thereby giving the product functional food status. Similarly, Valente et al. (2007b) studied CLA accumulation in rainbow trout (Oncorhynchus mykiss). Levels of 4.2% were achieved in the flesh. Ramos et al. (2008) studied the time course deposition of CLA in market size rainbow 16 trout (Oncorhynchus mykiss). Feeding the trout with 1% CLA during an 8-week period was sufficient to attain muscle CLA levels similar to those obtained after 12 weeks. A fast accumulation was registered reaching 1.3% of total fatty acids after just 2-week supplementation.

- Supplementing foods and diets with seafood-derived bioactives: The EPA and DHA content of daily diets in the USA can be increased significantly and cost effectively using nutraceuticals, functional and whole foods, and in this way meet the strategies of the American Heart Association in relation to secondary CHD prevention (Patterson and Stark, 2008). A good way to raise the ω- 3 PUFA content of the diet, without radical changes in eating habits, is to enrich commonly consumed food products with ω-3 PUFAs using fish oil preparations (30% EPA+DHA) and/or powder encapsulated forms (10% EPA+DHA) without significant flavour deterioration (Kolanowski et al., 1999). ω-3 PUFAs are being added to a plethora of diets and foods including eggs, breads, sausages, milk, reduced-fat spreads and others (Kolanowski and Laufenberg, 2006). However, the long-term supply of marine oil, the high energy cost of obtaining it (e.g. diesel for trawlers), and the lack of non-marine, synthetic or biosynthetic sources of ω-3 PUFAs raises serious sustainability issues and a likely shortage even in the short term (Gormley, 2010b). Picone et al. (2009) commented on the strong mediatic pressure for supplementing diets of pregnant women with ω-3 PUFAs and suggest that apart from a positive effect on cerebral development, the overall evidence for supplementation is not yet strong.

Dairy

There is no doubt that dairy products are functional foods. They are one of the best sources of calcium, an essential nutrient which can prevent osteoporosis and possibly colon cancer. In view of the former, the National Academy of Sciences recently increased recommendations for this nutrient for most age groups. In addition to calcium, however, recent research has focused specifically on other components in dairy products, particularly fermented dairy products (probiotics).

It is estimated that over 400 species of bacteria, separated into two broad categories, inhabit the human gastrointestinal tract. The categories are: those considered to be beneficial (e.g., Bifidobacterium and Lactobacillus and those considered detrimental (e.g., Enterobacteriaceae and Clostridium spp.). Of the beneficial microorganisms traditionally used in food fermentation, lactic acid bacteria have attracted the most attention (Sanders, 1994). Although a variety of health benefits have been attributed to probiotics, their anticarcinogenic, hypocholesterolemic and antagonistic actions against enteric pathogens and other intestinal organisms have received the most attention (Mital and Garg, 1995).

The hypocholesterolemic effect of fermented milk was discovered more than 30 years ago during studies conducted in Maasai tribesmen in Africa (Mann et al., 1964). The Maasai have low levels of serum cholesterol and clinical coronary heart disease despite a high meat diet. However, they consume daily 4 to 5 liters of fermented whole milk. Although a number of human clinical studies have assessed the cholesterol-lowering effects of fermented milk products (Sanders, 1994), results are equivocal. Study outcomes have been complicated by inadequate sample sizes, failure to control nutrient intake and energy expenditure, and variations in baseline blood lipids.

More evidence supports the role of probiotics in cancer risk reduction, particularly colon cancer (Mital and Garg, 1995). This observation may be due to the fact that lactic acid cultures can alter the activity of fecal enzymes (e.g., b-glucuronidase, azoreductase, nitroreductase) that are thought to play a role in the development of colon cancer. Relatively less attention has been focused on the consumption of fermented milk products and breast cancer risk, although an inverse relationship has been observed in some studies (Talamini et al., 1984; van't Veer et al., 1989).

Table 12: Probiotics

Probiotics at a glance

Probiotics are defined as live microorganisms – mostly bacteria – which when taken in adequate amounts confer a health benefit. Foods containing probiotic bacteria have been consumed for thousands of years and over time have developed a reputation for promoting health.

Traditionally, bacteria have been used for the production of fermented foods such as yogurt and sauerkraut. However, most of the bacterial strains that have been used for many centuries don't survive well in the gut – which is vital for any beneficial effect on health to take place. Specific bacterial strains (mainly Lactobacillus and Bifidobacterium species) have been found to be more resistant and are more able to survive passage through the gut into the large bowel. Many of these probiotic bacterial strains are now added to foods (e.g. dairy products), or are available as supplements in the form of capsules or sachets.

Only when a sufficient number of bacteria survive this journey, and are able to thrive and grow in numbers in the large bowel, can they begin to influence the composition of the intestinal flora. This is the name for the very large numbers of bacteria that live naturally within the large bowel, and the intestinal flora has relatively recently been recognised to contribute to our health and wellbeing in a number of ways. The types of bacteria present in the large bowel can influence our health both positively and negatively.

There is growing evidence that a regular intake of probiotics may positively influence our health. Their way of acting on our body is very complex and so not all the benefits for our health are well established, and further research in this field is required. Yet, some studies have delivered promising results. Probiotics seem to positively influence digestive health.

In addition to probiotics, there is growing interest in fermentable carbohydrates that feed the good microflora of the gut. These prebiotics, defined by Gibson and Roberfroid (1995) as "nondigestible food ingredients that beneficially affect the host by selectively stimulating the growth and/or activity of one or a limited number of bacteria in the colon and thus improves host health," may include starches, dietary fibers, other non-absorbable sugars, sugar alcohols, and oligosaccharides (Gibson et al., 1996). Of these, oligosaccharides have received the most attention, and numerous health benefits have been attributed to them (Tomomatsu, 1994). Oligosaccharides consist of short chain polysaccharides composed of three and 10 simple sugars linked together. They are found naturally in many fruits and vegetables (including banana, garlic, onions, milk, honey, artichokes). The prebiotic concept has been further extended to encompass the concept of synbiotics, a mixture of pro- and prebiotics (Gibson and Roberfroid, 1995). Many synbiotic products are currently on the market in Europe.

Beef

An anticarcinogenic fatty acid known as CLA was first isolated from grilled beef in 1987 (Ha et al., 1987). CLA refers to a mixture of positional and geometric isomers of linoleic acid (18:2 n-6) in which the double bonds are conjugated instead of existing in the typical methylene interrupted configuration. Nine different isomers of CLA have been reported as occurring naturally in food. CLA is unique in that it is found in highest concentrations in fat from ruminant animals (e.g., beef, dairy, and lamb). Beef fat contains 3.1 to 8.5 mg CLA/g fat with the 9-cis and 11-trans isomers contributing 57- 85% of the total CLA (Decker, 1995). Interestingly, CLA increases in foods that are cooked and/or otherwise processed. This is significant in view of the fact that many mutagens and carcinogens have been identified in cooked meats.

Over the past decade, CLA has been shown to be effective in suppressing forestomach tumors in mice, aberrant colonic crypt foci in rats, and mammary carcinogenesis in rats (Ip and Scimeca, 1997). In the mammary tumor model, CLA is an effective anticarcinogen in the range of 0.1-1% in the diet, which is higher than the estimated consumption of approximately 1 g CLA/person/day in the United States. These results are not due to displacement of linoleic acid in cells, suggesting that there may be unique mechanism(s) by which CLA modulates tumor development. Thus, there has been research designed to increase the CLA content in dairy cow milk through dietary modification (Kelly et al., 1998).

More recently, CLA has been investigated for its ability to change body com-position, suggesting a role as a weight-reduction agent. Mice fed CLA-supplemented diets (0.5%) exhibited 60% lower body fat and 14% increased lean body mass relative to controls (Park et al., 1997), possibly by reducing fat deposition and increasing lipolysis in adipocytes.

FOODS WITH ADDED FUNCTIONAL INGREDIENTS

As the products of scientific innovation and industrial processing, functionally enhanced foods promise healthfulness but conflict with understanding held by many consumers that healthy foods are those that are the most "natural." In interviews with consumers in five European cities, Holm et al. (2003) found that healthy eating means avoiding additives, and that eating "pure" foods (homemade from raw ingredients and containing few additives) was understood to provide protection against the dangers of "modern foods." Niva's subjects described healthy food as "pure," "natural," and "unprocessed" and felt that "foods enriched or fortified with healthy ingredients, even if they were extracted from nature, cannot attain the original balance and perfection of unprocessed foods" (Niva, 2007). Researchers report that while American consumers are receptive to the idea that some foods are naturally functional, they are skeptical of functionally engineered foods that emerge from labs and factories. In their 2007 telephone poll of 682 people, the Hartman Group found that when consumers are looking for health promotion and disease prevention they choose fresh and "real" foods (vegetables, fruits, vegetable juice, water, whole grains, fruit juice, fresh seafood, soy products, dairy products) over fortified or enhanced foods (Demeritt, 2008). 3e association of "naturalness" with healthfulness may seem irrational to those who are using technologies to improve the nutritional profile (and safety) of foods through processing, but it is consistent with long-standing beliefs associating health with a simpler, more rural past in which people lived closer to nature. 3e historian Rachel Lauden argues that the idea that food was better and healthier in the pre-industrial past is based on a distorted view of history. She reminds us that, for our ancestors, "Natural was something quite nasty" and describes a not-so-distant past in which natural not only tasted bad, but was also unreliable and often indigestible. Processed and preserved foods were healthier, tasted better and freed those who were not members of the aristocracy (in some cases up to 95% of the population) from unending agricultural and domestic toil (Lauden, 2001). Nonetheless, the idea that pre-modern foods were more pure, healthy and wholesome, and that contemporary eaters should avoid the dangers of processed foods by eating as much as possible like our great-great-grandmothers did, clearly has traction, as evidenced by the popularity of Michael Pollan's work advising exactly this (Pollan, 2007, 2008). In fact, longing for a past seen as simpler, more natural, and healthier is a historically consistent companion to our ongoing quest for progress through innovation and technology. 3ere is evidence that even Plato longed for a more wholesome past, a lost Golden Age that included whole meal ground at home (McCance and Widdowson, 1956). Foods offering nutritionally engineered functional benefits may appeal to our love of science, innovation and progress, but they also push up against these deeply held beliefs about the dangers of technology and the reassuring purity of those foods that are as close to nature as possible. Because the distinction between processed and "natural" foods is one that many consumers find important when choosing a healthy diet, it may be counterproductive to continue to include intrinsically (or naturally) functional foods (such as blueberries and almonds) and functionally enhanced foods in the same broad category of "functional foods." Gyorgy Scrinis (2008) argues that the inclusion of whole or unmodified foods in the category of functional foods is a sleight of hand that extends the "aura of healthfulness attached to whole foods across to the 'functional' processed foods" and undermines 104 Promoting Health by Linking Agriculture, Food, and Nutrition the "processed / unprocessed distinction as a basis for evaluating the healthfulness of foods." 3e Hartman Group has found that consumers do not use or understand the term "functional foods" in the same way that those in the food industry do; for lay people, the notion of functionality is integrally related to the natural health-giving properties in food. 3ey suggest that opportunity lies in offering products that mediate between the poles of naturally functional and artificially engineered foods (Demeritt, 2008). 3ose who are developing and promoting functional foods can better address consumers in their contexts by adopting terminology and practices that recognize the meaningfulness of the distinction between intrinsically and enhanced functional foods while striving to find the optimal and acceptable intersection between them.

Role of food as an agent for improving health has proposed a new class of food, called functional food, with positive effects on host health and/or well-being beyond their nutritional value. In the food industry, functional foods have been broadly defined as "foods similar in appearance to conventional foods that are consumed as part of a normal diet and have demonstrated physiological benefits and/or reduce the risk of chronic disease beyond basic nutritional functions."

Figure 17: Prominent types of functional foods.

The unique features of a 'functional food' are : "A conventional or everyday food, consumed as part of the normal/usual diet, composed of naturally occurring (as opposed to synthetic) components, perhaps in unnatural concentrations or present in foods that would not normally supply them, having a positive effect on target function(s) beyond nutritive value/basic nutrition, that may enhance well–being and health and/or reduce the risk of disease or provide health benefit so as to improve the quality of life including physical, psychological and behavioural performances and have authorized and scientifically based claims".

Fortified foods are foods that have nutrients added to them to boost the levels that are naturally present or to restore nutrients lost during processing. For example, a number of vitamins and minerals are added to white bread by law as they are removed when the wheat is milled to make flour. Some functional foods could also be considered to be fortified foods, and vice versa.

Bioactive compounds

The interest in functional foods continues to grow, powered by progressive research efforts to identify properties and potential applications of bioactive substances, and coupled with public interest and consumer demand. In the past decade, substantial progress has been made concerning our knowledge of bioactive components in plant foods and their links to health. Human diets of plant origin contain many hundreds of compounds which cannot be considered as nutrients, but appear to play a role in the maintenance of health. Evidence for the existence of bioactive compounds is based primarily on observational studies that demonstrate the beneficial effects of certain dietary patterns that include vegetarianism, high whole–grain consumption, the "prudent" diet, the Mediterranean diet, and the traditional Japanese diet. The traditional Japanese diet has a high content of soybean products and vegetables. The Mediterranean diet has a high content of olive oil, fruits and vegetables, and wholegrain breads. The "prudent" diet is characterized by high intakes of fruits and vegetables, fish, poultry, whole–grain products, and legumes. Many of the characteristic components of the traditional Mediterranean diet are known to have positive effects on health, capacity and well–being, and can be used to design functional foods. Vegetables, fruits and nuts are all rich in flavonoids, isoflavonoids, phytosterols and essential bioactive compounds providing health benefits. The polyunsaturated fatty acids found in fish effectively regulate haemostatic factors, protect against cardiac arrhythmias, cancer and hypertension, and play a vital role in the maintenance of neural functions and the prevention of certain psychiatric disorders.

Bioactive components include a range of chemical compounds with varying structures such as carotenoids, flavonoids, phytosterols, omega–3 fatty acids (n–3), allyl and diallyl sulfides, indoles (benzopyrroles), and polyphenols.

Bioavailability

Functional food research holds many promises for improving the quality of life for consumers; however, to achieve such outcomes, scientific research must effectively establish the bioavailability and efficacy of these compounds at levels that are physiologically achievable under typical dietary patterns. To date, numerous types of research have been employed to characterize functional compounds within food and their potential benefits.

Research designs used to investigate the bioavailability and efficacy of bioactive food compounds in food range from soft to hard science methodologies or from epidemiological studies to in vitro (test tube) studies using cell cultures or in vivo studies using animal models, and, ultimately, humans. Among primary research studies employed to elucidate the effects of bioactive components, each study design can be strengthened by employing the principles of sound science, which include appropriate test controls, blinding (single or double), and analysis of the bioactive compound or its byproduct of metabolism within the intended target organs or tissues. According to the FDA, bioavailability is defined as the rate and extent to which the active or therapeutic moieties of a compound are absorbed and become available at the intended site of action (Code of Federal Regulations, 2012). In essence, bioavailability is a measure of the absorption and availability for utilization of a specific substance. However overtly simple this may appear, bioavailability of a functional food compound (nutrient or nonnutrient) is governed by the physical properties of the food matrix and composition of the diet or mixed meal, as well as physiological differences among individuals in terms of absorption and metabolism of food. With regard to the food matrix or the vehicle housing the bioactive food compounds, the physical properties of the food matrix have been shown to affect the efficiency of the physical, enzymatic, and chemical digestion processes (Boyer and Liu, 2004). Once bioavailability is established, studies are employed to evaluate the efficacy of these compounds at producing desirable outcomes of interest in the intended target populations. Among the more challenging aspects of functional food research is substantiation of the science behind compound functionality by providing the compound at levels that are physiologically achievable within a food system or matrix. As such, human intervention trials become paramount to the advancement of functional food research for several reasons, including the fact that the whole food or whole formulated functional foods cannot be provided in cell culture. To date, numerous bioactive compounds have been isolated and evaluated using in vitro models to elucidate functional properties; however, the primary concern arising from use of in vitro models only is that the concentration of the individual compounds given to elicit an effect is typically outside of the context of normal in vivo dietary doses. Not only are these intake levels not achievable under most dietary patterns, but when supplied outside of a food matrix, unrealistic results may be touted and other more likely interactions may be overshadowed or can cause harm, as in the case of several supplementation trials (The Alpha-Tocopherol, Beta-Carotene Cancer Prevention Study Group, 1994; Leppala et al., 2000; Bjelakivik et al., 2007). To further strengthen the position of food as the vehicle for functional bioactive compounds, research has shown that some compounds can reach a point of saturation or toxic levels within the body, which limits any potential benefit (Eichhorn et al., 2004). For example, under conditions of excessive intake, bioactive compounds with antioxidant properties can shift the antioxidant oxidant balance within the body, such that antioxidants can behave as pro-oxidants and contribute to oxidative stress (Stanner et al., 2004; Yusuf et al., 2000; Brown et al., 2001; Moher et al., 2009). For these reasons, the field of functional food research must aid in establishing appropriate intake levels for these potentially beneficial compounds.

Table 13: Functional foods categories

Category	Definition	Examples
Basic foods	• The food or food product naturally contains the bioactive.	• Carrots naturally contain beta-carotene. • Oat bran cereals naturally contain beta-glucan.
Processed foods with added bioactives	• The bioactive does not exist naturally in the food and is added during processing.	• Orange juice with added calcium. • Milk with added omega-3 fatty acids.
Foods enhanced to have more of a bioactive	• The bioactive exists naturally in the food but the level of the bioactive is modified or concentrated (e.g. by traditional breeding, special livestock feeding or genetic engineering).	• Yogurt with increased levels of probiotics. • Tomatoes with increased levels of lycopene. • Eggs with increased levels of omega-3 fatty acids.

Food matrix

The food form refers to the matrix or vehicle in which the functional food exists. As described by Health Canada, a food form is "a format and serving size consistent with food use" (Health Canada, 2010), as opposed to a pill, tablet or capsule. There are many food forms in which functional foods can exist, some of which include beverages, yogurt, margarine, milk, juice and cereal. The form or matrix in which the functional food exists is relevant to consider from multiple perspectives, one of which includes consumer acceptance. Research indicates that the food form can influence consumer acceptance of functional foods. Dean et al. (2007) investigated Europeans' perception of functional foods, where the bioactive was incorporated into different grain products. Consumers viewed functional foods in "staple" food forms (e.g. bread) more favourably than functional foods in "fun" food forms (e.g. cookies) (Dean et al., 2007). Van Kleef et al. (2005) surveyed Dutch consumers and found functional foods were viewed more positively if they were in a food form perceived as healthy (e.g. yogurt, margarine, brown bread)9. Although not specific to functional foods, Balasubramanion and Cole (2002) found consumers paid more attention to the nutritional information of foods they perceived as healthy compared to foods they perceived as unhealthy (Balasubramanian, 2002). Bech-Larsen and Grunert (2003) found that consumers perceive a greater benefit in adding bioactives to food forms perceived as unhealthy, as opposed to adding bioactives to food forms perceived as already healthy (Bech-Larsen, 2003). Ares and Gambaro (2007) found that the combination of the food form and constituent bioactive can influence consumers' perceptions of healthiness of and willingness to try functional foods (e.g. yogurt with calcium was rated more positively than yogurt with antioxidants). These results were confirmed by Ares et al. (2008), suggesting that both the food form and bioactive contribute toward consumer acceptance. Krutulyte et al. (2008) focused specifically on omega-3 fatty acids in their interview of Dutch participants and found that addition of omega-3 fatty acids to fish balls, tuna salad and rye bread was perceived to be more appropriate compared to yogurt or sports nutrition bars14. These studies provide evidence that the functional food form can influence consumer acceptance. Perceptions of food forms as 'healthy' or 'unhealthy' is clearly a factor; however, there is also the consideration of the bioactive, the health benefit of interest and the interactions among them that will collectively influence consumer acceptance of the functional food. In addition to influence on consumer perception, it is also important to consider the food form in relation to its influence on the bioavailability and effectiveness of the constituent bioactive (FAO, 2007). FAO documents that the effect of the food form on the bioactive it contains must be considered during food processing, storage, preparation and consumption (FAO, 2007). If the food form can maintain the activity of the bioactive until the time of consumption and deliver the bioactive to the target area after consumption, it is an appropriate food form for the functional food. An example of this can be found in a meta-analysis by AbuMweis et al. (2008) that focused on the effectiveness of plant sterols in reducing LDL cholesterol and found that they are most effective when incorporated into foods containing fats. Specifically, plant sterols incorporated into fat spreads, mayonnaise, salad dressings, milk and yogurt more effectively lowered LDL cholesterol compared to plant sterols incorporated into croissants, muffins, orange juice, non-fat beverages, cereal bars and chocolate (AbuMweis et al., 2008).

Functional Foods Added with Essential Fatty Acids

For over 80 years the importance of PUFA consumption for human health has been established. The FDA recently approved the use of ω-3 PUFAs in supplements. Additionally, the market for ω-3 PUFA ingredients grew by 24.3% last year, which affirms their popularity and public awareness of their benefits. PUFAs are essential for normal human growth; however, only minor quantities of the beneficial ω-3 PUFAs, EPA and DHA are synthesized by human metabolism. Rather PUFAs are obtained via dietary or nutritional supplementation and modified into other beneficial metabolites. A vast literature base is available on the health benefits and biological roles of ω-3 PUFAs and their metabolism; however, information on their dietary sources and palatability of foods incorporated with ω-3 PUFAs is limited. DHA and EPA are added to many foods that are commercially available, such as infant and pet formulae, and they are also supplemented in animal feed to incorporate them in consumer dairy, meat, and poultry products. The chief sources of EPA and DHA are fish oils or purified preparations from microalgae, which when added to foods, impart a fishy flavor that is considered unacceptable. This fishy flavor is completely eliminated by extensively purifying preparations of n-3 PUFA sources. While n-3 PUFA lipid autoxidation is considered the main cause of fishy flavor, the individual oxidation products identified thus far, such as unsaturated carbonyls, do not appear to contribute to fishy flavor or odor. Alternatively, various compound classes such as free fatty acids and volatile sulfur compounds are known to impart fishy flavor to foods. Identification of the causative compounds to reduce and eventually eliminate fishy flavor is important for consumer acceptance of PUFA-fortified foods.

With continued experimentation, workers in the field have begun to understand the nutritional need for fats and in particular fatty acids in the diet. For this reason, many in the food industry have begun to focus on fatty acids and lipid technology as a new focus for food production, with its consequent benefits for the consumers consuming the modified products. This focus has been particularly intense for the production and incorporation of omega-3 fatty acids into the diet. Omega-3 fatty acids are long-chain polyunsaturated fatty acids (18-22 carbon atoms in chain length) (LC-PUFAs) with the first of the double bonds ("unsaturations") beginning with the third carbon atom from the methyl end of the molecule. They are called "polyunsaturated" because their molecules have two or more double bonds "unsaturations" in their carbohydrate chain. They are termed "long-chain" fatty acids since their carbon backbone has at least 18 carbon atoms. In addition to stearidonic acid "SDA" the omega-3 family of fatty acids includes alpha-linolenic acid ("ALA"), eicosatetraenoic acid (ETA), eicosapentaenoic acid ("EPA"), docosapentaenoic acid (DPA), and docosahexaenoic acid ("DHA"). ALA can be considered a "base" omega-3 fatty acid, from which EPA and DHA are made in the body through a series of enzymatic reactions, including the production of SDA. Most nutritionists point to DHA and EPA as the most physiologically important of the omega-3 fatty acids with the most beneficial effects.

Both omega-3 and omega-6 fatty acids are considered essential because humans, and other animals, cannot synthesize them and must instead get them from the diet. Western diets are plentiful in sources of omega-6 fatty acids (corn oil, palm oil and most nuts) but lacking in omega-3s (green leafy vegetables, flaxseed, olive oil, fish). To put the imbalance in perspective, it is estimated that Western diets contain between 10 and 15 times more omega-6 fats than omega-3 fatty acids. The "optimal" ratio of omega-6 to omega-3 is estimated to be closer to 2:1 (Benbrook et al., 2013).

The issue is not that omega-6 fatty acids are unhealthy; as LCPUFA they are considered one of the "good" fats and an important part of a healthy diet. But too many omega-6s can interfere with the way the body utilizes omega-3 fatty acids and thereby limit their numerous health benefits, such as reducing the risk of cardiovascular disease, diabetes and obesity. The prevalence of omega-6s in the Western diet may make it challenging for many people to reduce their consumption.

Eggs

Eggs are an exceptionally nutritious, high-quality protein food, but one concern lies in the fact that Chicken eggs have been touted as high in cholesterol, which contributes to hardening and narrowing of the arteries, heart attack, stroke, blocked blood vessels in the legs and kidneys, and may even contribute to dementia. But new studies have been done disproving that eggs and their yolks are bad for us, and now with Omega-3 enriched eggs, the yolk can actually help lower cholesterol and offer us many more health benefits.

Eggs can be fortified with two different omega-3 fatty acids: DHA (docosahexaenoic acid) and ALA (alpha linolenic acid). DHA is found in oily fish like salmon, trout and sardines. It's crucial for the proper development and maintenance of brain cells. Higher intakes of DHA and EPA (eicosapentaenoic acid), the other omega-3 fatty acid in oily fish, are also thought to guard against heart disease.

ALA is plentiful in flaxseed, flax oil, chia seeds, hemp oil, walnuts and walnut oil. It's also found in canola oil and soybeans. Some studies suggest that higher intakes of ALA are protective against heart disease. However, the evidence for ALA's heart benefits is less convincing since it hasn't been studied as extensively as DHA.

Omega-3 eggs are produced by hens fed a diet containing flaxseed. When the hens digest the flax, some of the ALA gets broken down into DHA and both fatty acids transfer to the yolk. One omega-3 egg typically contains 340 milligrams of ALA and 75 to 100 milligrams of DHA. Some companies also add fish oil to the chicken's feed to further increase the DHA content of egg yolks. GoldEgg Omega Choice, for example, provides 130 mg of DHA an egg.

There is no official recommended intake for DHA and EPA. Many experts recommend a daily intake of 1,000 mg of DHA + EPA (combined) for heart health. Eating an omega-3 egg each day will deliver only a fraction of this amount.Nevertheless, the advantages of simultaneous enrichment of eggs with vitamin E, carotenoids, selenium and DHA include better stability of polyunsaturated fatty acids (PUFA) during egg storage and cooking, high availability of such nutrients as vitamin E and carotenoids, absence of off-taste and an improved anti-oxidant and n-3 status of people consuming these eggs. Having reviewed the relevant scientific literature, "designer eggs" can be considered a viable type of functional food.

Table 14: Selected fatty acid content of vegetable oils with modified fatty acid profiles compared with the commodity oil

OIL Total	Oleic (18:1)	Linoleic (18:2)	Alpha-linolenic (18:3)	Saturates
CANOLA				
Conventional	60	20	10	7
High oleic	75	14	3	<7
High oleic	84	5	3	5
Low linolenic	65	22	4	7
Low linolenic P6	78	11-13	2-3	N/A
High myristate/palmitate	34	15	4	43
High laurate (37%)	34	12	7	45
SUNFLOWER				
Conventional	20	65	<1	10
High oleic	82	10	<1	8
Mid oleic	56	33	<1	9

SOYBEAN				
Conventional soy	23	51	7	14
Low linolenic	23	60	2	15
High palmitic (17%)	17	55	8	20
High stearic (28%)	20	35	7	35
High oleic soybean	83	2	3	12
OTHERS				
Conventional safflower	14	75	0	6
High oleic safflower	75	14	0	6
Conventional corn	24	58	<1	13
High oleic corn	70	-	-	-
Olive	75	8	<1	14
Avocado	65	15	1	14

Adapted from Pew Foundation 2007

Peanutbutter

Peanut is an important crop grown worldwide. Commercially it is used mainly for oil production but apart from oil, the by-products of peanuts contain many other functional compounds like proteins, fibers, polyphenols, antioxidants, vitamins and minerals which can be added as a functional ingredient into many processed foods. Recently it has also revealed that peanuts are excellent source of compounds like resveratrol, phenolic acids, flavonoids and phytosterols that block the absorption of cholesterol from diet. It is also a good source of Co-enzyme Q10 and contains all the 20 amino acids with highest amount of arginine. These bioactive compounds have been recognized for having disease preventive properties and are thought to promote longevity. The processing methods like roasting and boiling have shown increase in the concentration of these bioactive compounds (Arya et al., 2016). The high nutritional value and good taste make nut butter formulations, particularly peanut butter, an important food product in many households. New methods to improve the consumer appeal of peanut butters in respect to nutritional value, taste, and texture are continually being sought.

According to the American peanut council, peanut fat profile contains about 50 % monounsaturated fatty acids (MUFAs), 33 % Paraformaldehyde (PFAs) and 14 % saturated fatty acids which is a heart friendly combination of fatty acids (Feldman 1999). The amount of trans fat in peanut butter with 2 % stabilizer is 156 times less than what is needed to reach the 0 g trans-fat cut-off on food labels (Sanders 2001). Methods of making peanut butter spread formulations contain higher than naturally occurring levels of essential fatty acids are under development. These methods may include mixing ground, roasted nuts with omega 3-enriched oil to form an omega 3-enriched nut product; milling the omega 3-enriched nut product to produce a peanut butter formulation; and cooling the peanut butter formulation. In some products, at least one additional component, such as stabilizers, sweeteners, flavoring agents, bulking agents, fillings, and particulates may be added to the product.

Peanut products (raw, butter and oil) are more beneficial to heart health when compared to the low fat diets. The high monounsaturated fat peanut diets lowered their total body cholesterol by 11 % and bad LDL cholesterol by 14 %, while their good HDL cholesterol was maintained with reduction in triglycerides (Pelkman 2004). The benefits of the peanut diets on cholesterol were comparable to the olive oil diet. There is strong evidence supporting an association between monounsaturated fat as well as overall nut intake and reduction in the risk of coronary heart disease (Matilsky et al. 2009). Emerging data clearly shows that type of fat can impact health in various ways at different stages of life. The fat in peanuts and peanut butter provides healthy calories to malnourished infants and children at their time of need.

Peanuts are actually a legume and have more protein than any other nut with levels comparable to or better than serving of beans. After the peanut oil is extracted, the protein content in the cake can reach 50 % (Zhao et al. 2011). Peanuts contain all the 20 amino acids in variable proportions and is the biggest source of the protein called "arginine" (USDA 2014). According to Protein Digestibility Corrected Amino Acid Score (PDCAAS) peanut proteins and other legume proteins such as soy proteins are nutritionally equivalent to meat and eggs for human growth and health (FAO 2002). The amino acid profile of the peanut meals shows that it can be an ingredient for protein fortification (Yu et al. 2006). Since the proteins in peanuts is plant based, it carries with it additional components that have positive health benefits like fiber and unique bioactive components, unlike animal protein. The peanut proteins have been found to have good emulsifying activity, emulsifying stability, foaming capacity, excellent water retention and high solubility, and can also provide a new high protein food ingredient product formulation and protein formulation in food industry (Wu et al. 2009). Based on these observations, recently peanut protein has been incorporated into noodles (Wu et al. 2007) and infant formula, (Nimsate et al. 2010).

Milk

In the early 20th century, vitamin D was added to commercial cow's milk in response to the rise in malnourished children and adults with insufficient amounts of this essential nutrient in their diets. Today, many Americans and other populations consuming a primarily Western diet face another nutritional challenge. Despite having plentiful amounts of fat, the Western diet is lacking in a specific group of fatty acids called omega-3s, touted for their benefits to heart and brain health. In an effort to increase omega-3 intake, food manufacturers have started fortifying commonly consumed foods, including breads, cereals and eggs, with these essential fatty acids. Cow's milk also is getting in on the act, in more ways than one. Although milk can be fortified with plant-based omega-3 fats, researchers are exploring the potential benefits of fortifying dairy cattle diets with omega-3s to influence the proportion of these healthy fats in milk.

Milk and dairy products have been associated with health benefits for many years due to their content of bioactive peptides, probiotic bacteria, antioxidants, vitamins, specific proteins, oligosaccharides, organic acids, highly absorbable calcium, conjugated linoleic acid and other biologically active components with an array of bioactivities: modulating digestive and gastrointestinal functions, haemodynamics, controlling probiotic microbial growth and immunoregulation. Consumer's increasing interest for maintaining or improving their health by eating these specific food products has led to the development of many new functional dairy products.

For years, organic producers have claimed their milk is nutritionally superior to regular milk. Specifically, they say that because their cows spend a lot more time out on pasture, munching on grasses and legumes rich in omega-3 fatty acids, the animals' milk is higher in these healthy fats, which are linked to a reduced risk of heart disease. But the evidence for this has been scant, except for some small studies from Europe.

Now, a new study evaluating organic milk produced in the U.S. finds that organic milk has about 62 percent more omega-3s, compared to milk produced by cows on conventional dairy farms. Cows raised on conventional farms typically spend a lot more time in a barn or confined, and instead of grazing, they are fed a diet of animal feed that contains a lot of corn.

When researchers polled consumers to learn what nutrients consumers believe their diets lack, omega-3 was their top response. There is a recognized deficiency of omega-3 in the typical American diet, with health organizations around the globe recommending that we consume more omega-3. With omega-3 enriched milk, it is possible to complement other nutrients found in pasture-based organic milk - mainly ALA omega-3 - with DHA and EPA, two other omega-3 fatty acids sourced from high-quality fish oil, to deliver a product that offers higher content of omega-3 nutrition.

Most dairy products that have been added with omega 3s do so through fish oil sourced from EPA/DHA-rich anchovies and sardines. Any impurities that may be present within the fish oil, such as mercury, are removed before the fish oil is concentrated. The end result is a relatively safe source of omega-3 fatty acids. Another way to enrich the milk is by way of animal feed composition. The omega-3 composition of milk can be influenced by the cow's diet – if cows eat a diet high in omega-3 fatty acids, then a small fraction of them "bypass" biohydrogenation by the rumen microbes, and so make it to the true stomach. Hence, feeding cows lush omega-3 rich green grass, rather than the drought-stricken pastures more commonly offered, is associated with an increase in omega-3 fatty acids. Nevertheless, this increase is not always significant. The high content of polyunsaturated fatty acids (especially ALA) in grasses at early stage of maturity is the main reason for the positive effects of pasture on the fatty acid composition of milk.

Orange juice

Omega-3 fatty acids are increasingly being added to orange juice, often via fish oil or flax seeds. Because the main source of omega 3 foods in the Western diet is fish, and because most people do not consume fish often enough to meet dietary requirements of omega 3s, fortified juice could offer an alternative to fish for boosting levels of healthy omega-3 fats, especially in children. With the exception of breaded fish sticks, which are not made from omega-3-rich fatty fish, children tend to shun fish.

Researchers found that orange juice fortified with the omega-3 fatty acid DHA was able to raise blood levels of DHA in 31, 4- to 12-year-olds who drank the juice for six weeks. The findings indicate that enriched juice may be an effective way to deliver the fatty acid. The remaining question is whether those increases translate into health benefits.

Globally, the market for omega 3 enriched drinks is growing steadily, especially in North America, according to a report from Zenith International Ltd, Bath, U.K., which claims the omega 3 enriched beverage market generated global sales in the region of over $7 billion during 2006. North America accounted for nearly 33% of global volume in 2006, followed by Western Europe at 24% and Asia/Australasia at 23% (Cosgrove, 2008).

Over the last few years, scientists and manufacturers have overcome many of the challenges that previously limited omega 3s' application in beverage products. Technology advancements, such as microencapsulation, have been major contributors to the uptake in omega 3 beverages because they help reduce the fishy odours and taste that tend to arise as omega 3 oxidizes. Providing improved solubility, enhancing bioavailability and particle size reduction to the nanometer level of water-insoluble lipophilic compounds allows for the development of more stable and customer-accepted products for the market.

As is the case with fats in general, PUFAs such as omega 3s are water-insoluble compounds. When mixed with known emulsifiers in water, microemulsions are typically formed. These have the disadvantage of being opaque, as the droplet size is in the order of a micron, a thousandth of a millimeter. Moreover, they are meta-stable, meaning they can readily break down and separate into oil and water phases. When PUFAs are appropriately mixed however, omega 3 oils readily disperse in water resulting in translucent, stable aqueous solutions even at very high concentrations. Clarity is achieved through particle size reduction down to the nanometer level (one billionth of a meter).

Circumventing the fishy odor and taste boils down to converting the nutrient into a more stable, water-soluble powder, which helps delay oxidation—the point at which fishy odours and tastes become apparent. There are many methods in the industry known to prolong sensory stability of PUFA microemulsions, but all have only limited success. These include use of antioxidants, enzymes, complexing agents, emulsion surface engineering, chemical derivatization, encapsulation and flavour masking. Alternatively, particle size reduction dramatically changes the characteristics of the solutions they provide.

In addition, water-soluble dried omega 3 powders inherently enhance the product's stability. When oils are absorbed, under solid support, they have a much lesser propensity to oxidize.It adds another aspect to stability which doesn't relate to the liquid (emulsion) phase but is an additional way to reformulate in order to promote long term stability and avoid fishy odor and taste.

The value of these new products for the development of consumer acceptance applications has been further enhanced by obtaining GRAS status for PTS (polyoxyethanyl-alpha-tocopheryl sebacate), the lead solubilizing compound of many omega-3 enriched drinks.

A key advantage for the future growth of beverages formulated to contain omega 3 is that omega 3 itself is suited for both genders and for consumers of all ages. DHA is a fundamental building block of brain and nerve tissue as well as the retina of the eye. It has proven to be particularly crucial in this regard to fetal development during the last trimester of pregnancy. Following birth, DHA is transferred to babies via breast milk. It is therefore important for expectant mothers to take on enough DHA not just on behalf of their children but also in order to maintain their own supplies. DHA deficiencies at this stage can hinder growth and cognitive development for the child and are thought to contribute to the risk of post-natal depression in the mother. For teenagers, although the marketing toward this group is far less specific than for the infant/child/mother market, there are several 'young' products on the market that might appeal to this audience thanks to their packaging and placement.

If the current trends continue, the future for omega 3 beverages is most certainly bright. Omega 3 is a key route for functional drinks to move into the consumer mainstream. Still a niche sector in relative terms, it holds considerable potential for further development, with a wide range of applications open across the food and drinks spectrum (Cosgrove, 2008).

Table 15: Highlights of the importance of polyunsaturated fatty acids

Facts about polyunsaturated fats
• Polyunsaturated fats are essential for life. The body cannot produce them, they must enter the body through food.
• The most important polyunsaturated fats are considered to be omega-3 and omega-6.
• Omega-3 is found in oily fish such as salmon, mackerel, herring and sardines, some seaweed, walnuts, canola oil and flaxseed.
• Omega-6 is found in corn oil, sunflower oil and rapeseed oil.
• Omega-3 and Omega-6 have many different functions and are required in particular for the body to build and repair cells.
• Omega-3 and omega-6 affect the body in different ways. Yet in many cases omega-3 attenuates the effects of omega-6. For example, when it comes to blood clotting and the immune system. Both lipids interact and it is therefore important to get through the food right amount of both.

Functional Foods Added with Plant Sterols

Plant sterols are plant compounds with similar chemical structure and biological functions as cholesterol (Piironen et al., 2000). Plant sterols contain an extra methyl, ethyl group or double bond. The most abundant plant sterols are sitosterol, campesterol and stigmasterol (Moreau et al., 2002). The daily dietary intake of plant sterol is 160–400 mg among different populations (Ahrens and Boucher, 1978). However, in the earlier stages of human evolution, some 5–7 million years ago, plant sterol intake in Myocene diets would have been considerably higher, up to 1 g/d (Jenkins et al., 2003). Dietary sources include vegetable oils (especially unrefined oils), nuts, seeds and grains (Piironen et al., 2000). Absorption efficiency for plant sterols in humans is considerably less than that of cholesterol. Percent absorption of the former is 2–5% (Ostlund et al., 2002) versus 60% for the latter (Bosner et al. 1999). Consequently, blood levels of plant sterols in humans are only 0.1–0.14% of cholesterol levels (Miettinen et al., 1990). Due to their structural similarity to cholesterol, plant sterols were first and foremost studied for their cholesterol absorption inhibition properties. In addition to their cholesterol lowering effect, plant sterols may possess anti-cancer (Awad et al., 2003), anti-atherosclerosis (Moghadasian et al., 1997 and 1999), anti-inflammation (Bouic, 2001) and anti-oxidation activities (van Rensburg, 2000).

Plant sterols are efficacious when consumed in: oil: water emulsions (Gremaud et al., 2001 and 2002); water as lecithin micelles (Ostlund et al., 1999); yogurt (Volpe et al., 2001 and Mensink et al., 2002); low fat filled milks (Thomsen et al., 2004; Poteau et al., 2003; Richelle et al., 2004); chocolate (De Graaf et al., 2002); cereal; snack bars, breads, and beverages (Maki et al., 2003 and Nestel, 2001); and beef/hamburger (Matvienko et al., 2002 and Carr et al., 20002). Efficacy of a soy stanol-lecithin powder in reducing cholesterol absorption and LDL-C has been evaluated in a randomized, double-blind parallel study (Spilburg et al., 20003). The subjects who followed a Step I diet consumed soy stanol-lecithin powder in a beverage. The provided daily dose of plant stanols was 1.9 g. The reductions in blood cholesterol and LDL cholesterol were 10.1 and 14.4%, respectively. In another group of subjects, cholesterol absorption was measured using 625 mg stanols provided in beverage or egg whites. Stanol-lecithin reduced cholesterol absorption by 32.1% and 38.2 % when consumed in beverage and egg white, respectively. The reduction in LDL cholesterol reported, using the previous different vehicles, ranged between 7–14 %, which is close to the reduction in LDL cholesterol reported in studies that used fat spread as a vehicle for delivering plant sterol (Westrate and Meiji, 1998; Hallikainen et al., 2000; Homma et al., 2003).

Commercially, plant sterols are currently contained in bars (Logicol-Australia, Benecol-UK), vegetable oils (Ekona-Japan; NutraLease Canola Active-Israel), orange juice (Minute Maid Heart Wise containing Cargill CoroWise plant sterols), mayonnaises (Logicol-Australia), milk (Benecol-UK, Logicol-Australia, SereCol-Argentenia), yogurt (Logicol-Australia; Benecol-UK), yogurt drinks (Benecol), soy milk (Pacific Foods), meat and soups (Raisio-Finland), and green teas (Chol zero, Korea). Plant sterols are also being sold or developed mixed with other functional ingredients such as: fiber (Unilever Fruit D'or-France); healthy oils (Benecol Olive Spread-UK); non-absorbable diacylglycerol (Kao-ADM Econa Healthy Cooking Oil; Enzymotec MultOil Platform, ArteriCare products, Israel); almonds, soy protein and viscous fibers (Jenkins et al., 2003); and minerals (Tikkanen et al., 2001 and Vaskonene et al., 2002). There is also interest to combine plant sterols with antioxidants, such as flavonoids, quercetins, and catechin; and a spice mixture developed by Selako, and marketed as Flavomare in Scandanavia. It is only a matter of time before ingredients such as CLA are mixed with plant sterols in various vehicles.

Various manufacturers also sell plant sterols in supplement form, and there is interest to develop plant sterols as drugs. Plant sterols may also be combined with other drugs that lower cholesterol through different mechanisms of action, including statins and ezetimibe (Manhas and Farmer, 2004; Smart et al., 2004). Recent evidence suggests that patients who had previous acute coronary syndrome benefited from aggressive LDL lowering with statins to levels substantially below current target levels (Cannon et al., 2004 and Topol 2004). This finding provides enthusiasm for developing novel plant sterol-drug and drug-drug combined strategies to aggressively lower LDL cholesterol in some populations. Despite evidence that plant sterols can effectively reduce LDL cholesterol and inhibit cholesterol absorption in vehicles other than spread type vehicles, regulatory agencies have been slow to accept plant sterols in foods other than spreads in some countries such as the USA (Fed Regist, 2000) and Australia (ANZFA, 2001). Rigorous efforts underway by food companies and other highly respected organizations to allow claims for plant sterols in foods other than spreads.

Breakfast cereals and bread

As one of the most frequently consumed food products across age groups, breakfast cereal and bread are considered a viable and well accepted matrix for the delivery of nutrients. Work by Soderholm et al. (2010) reported on a double-blind, dietary intervention trial in which the subjects (n = 68) were randomized to receive a rye bread (9.3 g/d fiber) with added plant sterols (2 g/d) (active) or without (control). In the second phase of the study the amount of rye bread was doubled providing 18.6 g/d fiber and in the active group 4 g/d plant sterols. Compliance was monitored utilizing 3-day food diaries and a novel rye fiber-derived biomarker in plasma. Intake of rye bread enriched with 2 g/d of plant sterols during two weeks reduced significantly serum total and LDL cholesterol, apoB/apoA1 and total cholesterol/HDL cholesterol ratios by 5.1%, 8.1%, 8.3% and 7.2%, respectively, compared to controls. Correspondingly, the following two-week treatment with 4 g/d of plant sterols resulted in 6.5%, 10.4%, 5.5% and 3.7% difference compared to controls, being most pronounced for LDL (0.33 mmol/L). The treatments did not affect lipophilic (pro) vitamin levels. The study concluded that rye bread enriched with 2–4 g/d of nonesterified plant sterols beneficially modifies cardiovascular lipid risk factors in normocholesterolemic subjects compared to controls.

Epidemiological studies have clearly shown that whole-grain cereals can protect against obesity, diabetes, CVD and cancers. The specific effects of food structure (increased satiety, reduced transit time and glycaemic response), fibre (improved faecal bulking and satiety, viscosity and SCFA production, and/or reduced glycaemic response) and Mg (better glycaemic homeostasis through increased insulin secretion), together with the antioxidant and anti-carcinogenic properties of numerous bioactive compounds, especially those in the bran and germ (minerals, trace elements, vitamins, carotenoids, polyphenols and alkylresorcinols), are today well-recognised mechanisms in this protection. Recent findings, the exhaustive listing of bioactive compounds found in whole-grain wheat, their content in whole-grain, bran and germ fractions and their estimated bioavailability, have led to new hypotheses. The involvement of polyphenols in cell signalling and gene regulation, and of sulfur compounds, lignin and phytic acid should be considered in antioxidant protection. Whole-grain wheat is also a rich source of methyl donors and lipotropes (methionine, betaine, choline, inositol and folates) that may be involved in cardiovascular and/or hepatic protection, lipid metabolism and DNA methylation. Potential protective effects of bound phenolic acids within the colon, of the B-complex vitamins on the nervous system and mental health, of oligosaccharides as prebiotics, of compounds associated with skeleton health, and of other compounds such as α-linolenic acid, policosanol, melatonin, phytosterols and para-aminobenzoic acid also deserve to be studied in more depth (Fardet 2010).

Table 16: Main bioactive compounds in wheat

Bioactive compound	Whole-grain wheat	Wheat bran	Wheat germ
Linolenic acid (18:3n-3)	-‡	0.16	0.53
Sulfur compounds	0.5	0.7	1.2
Total free glutathione§	0.007	0.038	0.270
Dietary fibre‖	13.2	44.6	17.7
Lignins	1.9	5.6	1.5
Oligosaccharides¶	1.9	3.7	10.1
Phytic acid	0.9	4.2	1.8
Minerals and trace elements	1.12	3.39	2.51
Vitamins	0.0138	0.0398	0.0394
B Vitamins	0.0091	0.0303	0.0123
Vitamin E (tocopherols and tocotrienols)	0.0047	0.0095	0.0271
Carotenoids	0.00034	0.00072	-‡
Polyphenols	0.15	1.10	>0.37
Phenolic acids	0.11	1.07	>0.07
Flavonoids	0.037	0.028	0.300
Lignans	0.0004	0.0050	0.0005
Alkylresorcinol	0.07	0.27	-‡
Betaine	0.16	0.87	0.85
Total choline	0.12	0.17	0.24
Total free inositols (*myo*- and total *chiro*-inositols)	0.022	0.025	>0.011
Phytosterols	0.08	0.16	0.43
Policosanol + melatonin + *para*-aminobenzoic acid	0.00341	0.00290	>0.00186
Total	>15.4	51.5	>23.9
Subtotal (without dietary fibre)	>2.2	6.9	>6.2

*Mean percentages of bioactive compounds found in wheat bran, whole-grain wheat and wheat germ are calculated as follows: % = (minimum value+maximum value)/2. †Expressed as g/100 g food. ‡No data found. §Total free glutathione is given as glutathione equivalents = reduced glutathione+(oxidised glutathione × 2). ‖Dietary fibre content is measured according to the AOAC method as such or modified. ¶Oligosaccharides include fructans, raffinose and stachyose. Adapted from Fardet, 2010.

Yogurt and other dairy drinks

A recent controlled clinical trial has shown that the intake of plant sterols provided in low-fat and non-fat beverages did not affect lipid profiles in moderately hypercholesterolemic individuals (Jones et al., 2003). The finding of this study was contrary to the findings of other studies reporting that plant sterols were effective in reducing blood cholesterol even when incorporated into low-fat or non-fat foods (Volpe et al., 2001; Thomsen et al., 2004; Pouteau et al., 2003; Mensink et al., 2002; Nestel et al., 2001). This discrepancy may be related to the fact that the plant sterols must be added to the low or non-fat food matrix in such a way that the plant sterols solubilize, or remain as small crystals over time. In products such as milk, the milk fat globule membrane components may enhance the absorption of cholesterol (Pouteau et al., 2003; Gremaud et al., 2001 and 2002), but also aid in solubilisation. Pouteau et al. (2003) described a rapid filtration and detection method to quantify plant sterol crystals. The authors also used light scattering techniques to quantify the size of the crystals as a function of storage time of the milk products. The method of dispersion, processing, and use of emulsifiers, surfactants, and crystal habit modifiers will affect the success of plant sterols in non-spread vehicles (Thomsen et al., 2004; Pouteau et al., 2003; Richelle et al., 2004; Groux et al., 2003; von Bonsdorff-Nikande et al., 2003).

Table 17: Top food sources of naturally occurring and commercially available plant sterols

Many of the foods that you already enjoy are rich sources of plant sterols and stanols. These include:

- corn oil (0.13g/Tbsp)
- sunflower oil (0.1g/Tbsp)
- beans (0.07g/1/2 cup)
- corn (0.06g/1/2 cup)
- peanut butter (0.05g/2 Tbsp)
- olive oil (0.03 g/1 Tbsp)
- almonds (0.02g/1 oz.)
- oranges (0.02g/1 small orange)
- apples (0.01g/1 small apple)
- avocados (0.008g/1 oz)

Sterols and stanols are also available in commercial foods that have been fortified with them. These include:

- Benecol spread (0.85g/tbsp.)
- Healthy Heart Yogurt (0.4g/6 oz.)
- Lifetime lowfat cheese (0.65g/1 oz.)
- Minute Maid Premium Heartwise orange juice (0.04g/8 oz.)
- Nature Valley Healthy Heart Chewy Granola Bars (0.4g/bar)
- Orowheat whole grain bread (0.4g/1 1/2 slices)
- Take Control spread (1.7g/Tbsp)

Products available in the USA as of 2015.

A study by Volpe et al. (2001) aimed to assess the effect of consumption of a yoghurt-based drink enriched with 1–2 g plant sterols/d on serum lipids, transaminases, vitamins and hormone status in patients with primary moderate hypercholesterolaemia. Thirty patients were randomly assigned to one of two treatment groups: a low-fat low-lactose yoghurt-based drink enriched with 1 g plant sterol extracted from soyabean/d v. a low-fat low-lactose yoghurt, for a period of 4 weeks. After a 2-week wash-out period, patients were crossed over for an additional 4-week period. Second, after a 4-week wash-out period, eleven patients were treated with 2 g plant sterols/d in a second open part of the study for a period of 8 weeks. The yoghurt enriched with plant sterols significantly reduced, in a dose-dependent manner, serum total cholesterol and LDL cholesterol levels and LDL-cholesterol:HDL-cholesterol (P,0:001), whereas no changes were observed in HDL-cholesterol and triacylglycerol levels, either in the first or the second part of the study. There were only slight, not statistically significant, differences in serum transaminase, vitamin and hormone levels. The study concluded that a low-fat yoghurt-based drink moderately enriched with plant sterols may lower total cholesterol and LDL-cholesterol effectively in patients with primary moderate hypercholesterolaemia. Based on these type of studies, many yogurt and yogurt-like drinks have since been developed.

In 2010 Parmalat launched the first probiotic yogurt available in Canada that offers options for consumers trying to manage their cholesterol levels. Each 100 g of Astro BioBest Plant Sterols yogurt contains 50 per cent of the daily amount of plant sterols recommended for the Canadian population. In July of the same year, Health Canada confirmed that plant sterols can be added to mayonnaise, margarine, calorie-reduced margarine, yogurt and yogurt drinks, vegetable and fruit juices, salad dressing and unstandardized salad dressings and spreads. Health Canada has also issued a guidance regarding the use of health claims and the cholesterol lowing effect of plant sterols. With the development of new products, the goal is to be able to provide consumers with a simple, everyday food that's easily incorporated into their diet with the added bonus of a functional ingredient that specifically lowers cholesterol (Food in Canada, 2010).

Butter-like spread

Plant sterols have been shown to reduce serum lipid concentrations. The effectiveness is highly dependent on the physical state of the plant sterols. By means of a new crystallizing method, plant sterols can be added into dietary fats and oils homogeneously. Most clinical trials have been conducted using plant sterols or stanols added to spreads due to their relatively easy incorporation processes when compared to other food matrixes. As long as plant sterols are consumed with a meal to stimulate biliary flow, they can effectively lower LDL cholesterol on the background of various types of basal diets and food vehicles.

Hendriks et al. (1999) carried out a study to investigate the dose-response relationship between cholesterol lowering and three different, relatively low intake levels of plant sterols (0.83, 1.61, 3.24 g=d) from spreads. One hundred apparently healthy normocholesterolaemic and mildly hypercholesterolaemic volunteers participated, with each subject consuming four spreads, each for a period of 3.5 week. The study spreads included butter, a commercially available spread and three experimental spreads fortified with three different concentrations of plant sterols. The three relatively low dosages of plant sterols had a significant cholesterol lowering effect ranging from 4.9 ± 6.8%, 6.7 ± 9.9% and 6.5 ± 7.9%, for total, LDL-cholesterol and the LDL=HDL cholesterol ratio, respectively, without substantially affecting lipid soluble (pro)vitamins. No significant differences in cholesterol lowering effect between the three dosages of plant sterols could be detected. This study supports the consumption of about 1.6 g of plant sterols per day to affect plasma cholesterol concentrations without seriously affecting plasma carotenoid concentrations.

In another study (Sierksma et al., 1999), seventy-six healthy adult volunteers with an average age of 44 (sd 11) years, with baseline plasma total cholesterol levels below 8 mmol/l, received in a balanced, double-blind, crossover design, a total of three different table spreads for personal use for 9 weeks. Two spreads were fortified either with free (non-esterified) vegetable-oil sterols, mainly from soyabean oil (31 g sterol equivalents/kg; 0·8 g/d) or sheanut-oil sterols (133 g sterol equivalents/kg; 3·3 g/d). One spread was not fortified (control). Average intake of spread was 25 g/d for 3 weeks. None of the spreads induced changes in blood clinical chemistry or haematology. Plasma total- and LDL-cholesterol concentrations were statistically significantly reduced by 3·8% and 6% (both 0·19 mmol/l) respectively, for the spread enriched with free soyabean-oil sterols compared with the control spread. The spread enriched with sheanut-oil sterols did not lower plasma total- and LDL-cholesterol levels. None of the plant-sterol-enriched spreads affected plasma HDL-cholesterol concentrations. Plasma-lipid-standardized concentrations of α- plus β-carotene were not statistically significantly affected by the soyabean-oil sterol spread in contrast to lipid-standardized plasma lycopene levels which showed a statistically significant decrease (9·5%). These findings indicate that a daily intake of free soyabean-oil sterols as low as 0·8 g added to a spread is effective in lowering blood total- and LDL-cholesterol levels with limited effects on blood carotenoid levels. The lowering in total- and LDL-cholesterol blood levels due to consumption of the vegetable-oil-sterol-enriched spread may be helpful in reducing the risk of CHD for the population.

Table 18: Effects of free sterols and stanols on LDL cholesterol

Products	Vehicle	Subject	Condition: Starting mean TC mM	Daily Dose g Free sterol/ stanol Equivalents	#Servings/d	Outcome: Where possible, Placebo-adjusted % decrease LDL is given = % decrease with sterol (T_o vs T_f) % decrease with placebo (T_o vs T_f). C-abs also indicated
A2 C2	Spread	9	Normocholesterol emic	1.0 1.3		↓ C-abs 42.0% ↓ C-abs 33.0%
B2	Sunflower oil (Capsule)	3 M, 3 W	Hypercholesterole mic	1.5		↓ LDL 14.4%
A1 B1	Mayonnaise	24 M&W	6.5	0.96 1.0	Habitual usage	↓ TC 5.7% ↓ TC 2.4%
A1 B1 D1	Mayonnaise	22 M, 9 W	>6.0	1.0 1.0 1.2	Habitual usage	↓ LDL 6.2% ↓ LDL 5.1% ↓ LDL 7.7%
B1+step 1 diet	Capsules with sterols consumed with food	33 M	6.2	3.0 3.0 with 8 g cholestyramine 8 g of cholestyramine	3	↓ LDL 1.8%
A2	Spread	12 M	Normocholesterol emic	0.7		↓ LDL 6.1%
A2	olive oil	8 M, 8 W	3.7 M, 4.1 W	0.4 g/1000 Kcal	3	↓ LDL 2.8%
A1	Spread	22	5.0 minimum	1.6		↓ LDL 15 % Normocholesterolemics ↓ LDL 4.7% hypercholesterolmics
B1+prudent diet	Spread	32 M	6.7–6.8	1.7	3	↓ LDL 15%
B2 +soylecithin	Spread	4 M, 2 W	5.2	1.0-B2 powder 0.3–0.7-B2/lecithin micellar mix		↓ C-abs 11.0% ↓ C-abs 34.4% at 0.3 g ↓ C-abs 36.7% at 0.7 g
A2	Spread	39 M, 37 W	4.9–5.1	0.8	Habitual usage	↓ LDL 6%, ↓ TC 9%
B1	Liquid emulsion	12 M	6.9	3	3	↓ C-abs 40%
A1 B1 1:1 mixture A1:B1	Butter fat ± sterols	9 M, 6 W	6.5	1.8	3	↓ LDL 13.3%, ↓ C-abs 56.1% ↓ LDL 13.4%, ↓ C-abs 33.3% ↓ LDL 16.0%, ↓ C-abs 49.2%

A2	Vegetable oil, partly filled milk	18 M	7.0	1.8	2	↓ C-abs 39%
A2	Vegetable oil, partly filled milk	33 M	<5.2	2.2	2	↓ C-abs 51.1%
A2	Vegetable oil, partly filled milk	18 M, 51 F	7.0	1.2 or 1.6	2	↓ LDL 7.1% at 1.2 g ↓ LDL 9.6% at 1.6 g

Abbreviation: C-abs, cholesterol absorption; LDL-C, low density lipoprotein cholesterol; TC, total cholesterol; T_o and T_f refer to start and end points of the study; M, men; W, women; A, free plant sterol; B, free plant stanol; C, plant sterol ester; D, plant stanol ester; 1, tall sterol; 2, soy sterol; 3, shea sterol; 4, corn sterol; 5, rice sterol. Adapted from Berger et al., 2004.

Chip-like snacks

Corazonas Foods, one of the leading companies in the development of sterol-rich functional foods, produces snacks like oatmeal squares, tortilla chips and potato chips that are fortified with plant sterols. While the FDA recommends an individual daily consumption of at least 0.8 grams of plant sterols in order to keep cholesterol at a healthy level, this company claims to have their oatmeal squares containing 0.8 grams of plant sterols per serving and their tortilla chips contain 0.5 grams of plant sterols. The FDA recommends at least 1.3 grams of plant sterol esters per day for significant cholesterol-lowering effect, while the AHA states that at least 2 grams are required to achieve the maximum cholesterol-lowering effect (Brown, 2012). The science behind the enriched snacks appears sound, and the cholesterol reduction measured in human testing was in line with benefits seen from other sterol-enriched foods such as some margarines and salad dressings that have been marketed for years.

Hayes et al. (2004) described LDL cholesterol reductions averaging 15 percent among 10 people. Participants ate two 1-ounce servings of phytosterol-enriched tortilla chips daily for four weeks. The researchers reported no changes in the test participants' HDL cholesterol, and a 10 percent drop in total cholesterol. According to the study authors, the sterol-enriched chips worked by blocking the body's absorption of LDL as other foods were eaten along with the chips.

Consumer reaction has been mixed on the scores of often taste-challenged foods that are lower in fat, carbs or sugar. Food industry analysts believe sterol-enriched snacks may get just as tough a reception from the medical community and from consumers wishing to reduce cholesterol — even if the new treat tastes the same as regular snacks.

It has been a widely held belief that to obtain appreciable benefit from phytosterols, either plant sterols, stanols, or combinations thereof (including beta-sitosterol, beta-sitostanol, campesterol, campestanol, stigmasterol, stigmastanol, brassicasterol, brassicastanol, clionasterol and clionastanol; collectively termed phytosterol or phytosterols) for lowering plasma cholesterol, the phytosterol should be dissolved in an edible oil or other solvent so that it can enter micelles in the small intestine to inhibit the absorption of cholesterol. This belief has been supported by early research carried out in the 1950s through the 1970s showing that large doses of phytosterols in their solid form (coarse powders) were required to achieve meaningful decreases in plasma cholesterol levels. For example, in 1956, Faquhar et al. (1956) showed that doses of 12–18 g per day of beta sitosterol (provided in divided doses) were required to achieve a 15–20% lowering of serum cholesterol in males with atherosclerosis. In another study, 9 g per day of soybean-derived phytosterols were required to lower plasma cholesterol approximately 9% (Kucchodkar et al., 1976). In yet another study, 3–9 g per day of tall oil-derived phytosterols were required to lower plasma cholesterol approximately 12% (Lees et al., 1977). In a recent study, 1.7 g per day of finely powdered tall oil-derived phytosterols were sufficient to lower total plasma cholesterol by 9% and LDL-cholesterol by about 15% (Jones et al., 1999).

It has been generally appreciated that phytosterols such as alpha and beta sitosterol, stigmosterol, campesterol, and the corresponding saturated (chemically reduced or hydrogenated) "stanol" species, are insoluble in water, and only slightly soluble in edible oils. Accordingly, to promote the solubilization of phytosterols, and their efficacy in lowering plasma cholesterol, Goto et al. (2000) described the incorporation of at least 15% and as much as 70% by weight or more of a polyhydric alcohol/fatty acid ester (including glycerol fatty acid esters containing at least two esterified and at least one unesterified hydroxyl group such as diacylglycerols or diglycerides), into a fat. Between 1.2% and 4.7% by weight of phytosterol is incorporated into the polyhydric alcohol/fatty acid ester containing fat composition. The high proportion of diacylglycerol assures solubility or dispersal of the phytosterol to provide a cholesterol-lowering fat substitute.

Akashe et al. (1999) described a plant sterol/emulsifier complex that has a lower melting temperature than the plant sterol alone. The complex, e.g., a co-crystallized monoglyceride and plant sterol mixture, is said to facilitate incorporation of the sterol into food products without adversely affecting the texture of the food products. It has been widely believed that increasing the solubility of phytosterols in fat increases their bioavailability and reduces the dose required to achieve a specified degree of cholesterol reduction. Thus, Miettinen et al. (1996), described the preparation and use of the plant stanol, beta sitostanol, in the form of a fatty acid ester which is readily soluble in an edible oil, to reduce the serum cholesterol level in humans. This technology has been utilized in manufacturing the margarine product marketed under the tradename Benecol®.

HEALTH BENEFITS OF FUNCTIONAL FOODS

In modernsocieties, many individuals do not consume a balanced diet and this has led to an increase in the incidence of serious metabolic imbalances. These in turn increase the risk of chronic conditions such as obesity, type 2 diabetes, hypertension, food allergies and intolerances, and gastrointestinal and inflammatory disorders. Good nutrition is vital for health, optimal growth and development, and for the prevention of disease (Liang et al., 2005). It is now understood that nutrients and other substances obtained from a wide variety of foods promote health, maintain metabolic homeostasis, and fulfill energy requirements. Nevertheless, due to the great variation in human genetics and culture, there is no optimal diet to fit everyone. For this reason, it is so important for food and nutrition scientists to determine the relationship between an individual's metabolism and their food selection.

The gut is an obvious target for functional foods as it is the boundary between what is eaten and the reactions that take place in the body.

Key digestive physiological processes that can be regulated by dietary modifications include:

- satiety
- the rate and extent of macronutrient breakdown
- absorption from the small intestine
- metabolism
- the nature of the gut microbiota
- fermentation
- mucosal functioning
- bowel activity
- gut immune system

Over the past few decades, several large non-nutrient molecules have been identified from various foods. These include secondary metabolites in plants that are used for funtions such as defense or reproduction but are not, until now, considered essential nutrients. Up until recent times, these secondary metabolites (such as phytochemicals) were mostly ignored. Today, their potential metabolic effects are much more largely appreciated. For example, flavones have been found to protect against heart disease and soy-based estrogens against cancer. The bioactivity of thousands of non-nutrient compounds in foods such as red wine, coffee, fish, and fruits and vegetables have been investigated to determine their effects on animal biochemistry and metabolism and to identify those with potential health benefits.

Probiotics are usually defined as microbial food supplements that have beneficial effects on consumers, as described in an earier section of this book. Microbiologists first reported that the microflora of the gastrointestinal tracts of healthy individuals differed from that of diseased individuals, and the microflora found in the healthy individuals was termed probiotic. Most probiotics fall into the group of organisms known as lactic acid-producing bacteria, and these are normally consumed in the form of yogurt, fermented milks, or other fermented foods. Many of the functional food products that have been developed include probiotics.

How functional foods affect human health

In order to better understand how functional foods may affect humanhealth it is important to understand what happens when food is digested and absorbed by the GI tract. The human gut is a selective nutrient absorption system. It is also the most important signalling system and information exchange system within the body. It acts as a signal translator, a sensor for neuroendocrinology, and an immunological recognition and presentation system. The GI tract is also a complex information exchange system comprising a number of signaling networks involving gut cells. These cells in turn will reach all nooks in the organism via blood transportation. The bioactivity of functional foods in vivo may be due to their effects on such networks, but this raises the question of what signaling pathways are used by non-nutrients that cannot be absorbed by the gut. For these reasons, the understanding of how food works in the body is extremely complex. Not only must we consider the activity mechanisms of what is absorved, but also what happens to and by the non-nutrients that are not absorved. Table 19 below summarized some bioactive components found in some functional foods, the health benefits that they may provide and the strength of evidence available for these effects until now.

All claims linking the consumption of functional foods or food ingredients with any potential health outcomes require comprehensive scientific evidence. The Food and Drug Administration (FDA) outlined the criteria for "significant scientific agreement" in a guidance document released on December 22, 1999 (USDA, 1999). Nevertheles, there is a clear discrepancy between "emerging evidence" (characterized by in vitro or animal studies, uncontrolled human studies, and inconsistent epidemiological evidence) and "significant scientific agreement." To reach such agreement requires the support of a body of consistent, relevant evidence from well-designed clinical, epidemiologic and laboratory studies, and expert opinions from a body of independent scientists. Claims about the health benefits of functional foods should be based on sound scientific evidence, but too often only so-called "emerging evidence" is the basis for marketing some functional foods or their components.

Table 19: Strength of evidence for some functional foods in the US market

Functional food	Bioactive component	Health Benefit	Type of evidence	Strength of evidence	Recommended amount of frequency of intake	Regulatory
Fortified margarines	Plant sterol and stanol esters	Reduce total and LDL cholesterol	Clinical trials	Very strong	1.3 g/d for sterols 1.7 g/d for stanols	Health claim
Psyllium	Soluble fiber	Reduce total and LDL cholesterol	Clinical trials	Very strong	1 g/d	Health claim
Soy	Protein	Reduce total and LDL cholesterol	Clinical trials	Very strong	25 g/d	Health claim
Whole oat products	Glucan	Reduce total and LDL cholesterol	Clinical trials	Very strong	3 g/d	Health claim
Cranberry juice	Proanthocyanidins	Reduce urinary tract infections	Small number of clinical trials	Moderate	300 mL/d	Conventional food
Fatty fish	(n-3) Fatty acids	Reduce TG, reduce health disease cardiac deaths and fatal and non-fatal myocardial infarction	Clinical trials; epidemiological studies	Strong	2/wk	Qualified health claim for dietary
Garlic	Organosulfur compounds	Reduce total and LDL cholesterol	Clinical trials	Moderate	600-900 mg/d	Conventional food or dietary supplement
Green tea	Catechins	Reduce risk of certain types of cancer	Epidemiological	Weak to moderate	Unknown	Conventional food
Spinach, kale, collard greens	Lutein/zeaxanthin	Reduce risk of age-related macular degeneration	Epidemiological	Weak to moderate	6 mg/d	Conventional food dietary supplement
Tomatoes and processed tomato products	Lycopene	Reduce risk prostate cancer	Epidemiological	Weak to moderate	Daily	Conventional food
Lamb, turkey, beef, diary	CLA	Reduce breast cancer	In vivo and in vitro studies	Weak	Unknown	Conventional food
Cruciferous vegetables	Glucosinolates indoles	Reduce risk of certain types of cancer	Epidemiological	Weak	3 or more servings/wk	Conventional food
Fermented dairy Products	Probiotics	Support GI health, boos immunity	In vivo and In vitro studies, limited clinical data	Weak	Dairy	Conventional food or dietary supplement

1 Foods that have an FDA-approved health claim (sterol/stanol esters, oats, psyllium, soy) generally are supported by two dozen or more well-designed published clinical traits. For example, the soy health petition contained > 40 clinical trials, while there are only a few clinical trials on cranberry juice and urinary tract infections.

2 TG, triglycerides; CLA conjugated linoleic acid; GI gastrointestinal.

Adapted from Hassler, 2002

Cardiovascular Benefits

Cardiovascular disease (CVD) is a general term for conditions affecting the heart or blood vessels. It is usually associated with a build-up of fatty deposits inside the arteries – known as atherosclerosis – and an increased risk of blood clots. It can also be associated with damage to arteries in organs such as the brain, heart, kidneys and eyes. CVD is one of the main causes of death and disability around the world, but it can often largely be prevented with a healthy lifestyle.

Inflammation

Many people think of inflammation in terms of external signs, as in when swelling and bruising occurs. While it is true that these two processes are indeed inflammation processes, internal (usually unperceived) uncontrolled inflammation plays a role in almost every major disease including: cancer, heart disease, diabetes, Alzheimer's disease and even depression.

Inflammation occurs naturally in the body as a defence mechanism triggered by the immune system. However, when inflammation is prolonged and becomes a chronic state it can trigger various disease processes. Over time, inflammation becomes too much of a good thing. With chronic inflammation, the body is on high alert at all times. This prolonged state of emergency can cause lasting damage to the heart, brain and other organs. For example, when inflammatory cells prolong their stay inside blood vessels, they promote the buildup of dangerous plaque. The body sees this plaque as foreign and sends more of its first responders. As the plaque continues to build, the arteries can thicken, making a heart attack or stroke much more likely to occur.

Likewise, inflammation in the brain may play a role in Alzheimer's disease. For many years the brain was thought to be off-limits to inflammation because of the blood-brain barrier, a sort of built-in security system, but scientists have proved that immune cells can and do infiltrate the brain during times of distress. Their role in disease progression is not yet clear. Inflammation is an immune system response to environmental irritants, toxins, and infection. When the immune system is activated by one of these intruders, pro-inflammatory hormones signal the white blood cells to rush in and clean up the infected or damaged tissue. Once the invaders have been subdued, anti-inflammatory agents move in to begin the healing process. In a normal immune system, a natural balance exists between inflammation and the anti-inflammatory agents.

Several dietary guidelines exist with the aim of preventing or managing chronic inflammation. Further development of functional foods that address this condition would also benefit most the population.Despite the connection between inflammation and prevalent diseases, as well as the connection between diet and inflammation, diet is not always considered in response to inflammation. To move toward an anti-inflammatory diet and anti-inflammatory foods, it is necessary to primarily move away from the abundance of overly processed, unbalanced diets of the West and toward the ancient eating patterns of traditional diets. In general, these include plenty of fresh fruits and vegetables, little meat, certainly no chemicals or meat additives, and an abundance of healthy fats. In other words, a more natural, less processed diet.

To prevent and reduce inflammation, the aim is to consume an overall healthy diet. Making fruits and vegetables the center of meals, including such as blueberries, apples, and leafy greens. Papayas and cherries are right up there too. These are high in natural antioxidants and polyphenols as previously discussed. Nuts have also been associated with reduced markers of inflammation and a lower risk of cardiovascular disease and diabetes; as have other healthy sources of fat like avocados and oily fish. Coffee, which contains polyphenols and other anti-inflammatory compounds, may protect against inflammation as well. Food is one of the best inflammation fighters. The food that is eaten will either help feed inflammation or extinguish it.While the number one culprit when it comes to diet and inflammation has been identified as simple sugar, there are also several nutrients that may help fight off this chornic condition. Among them it is possible to highlight antioxidants and essential fatty acids.

PUFAs, notably EPA and DHA, are among 6 the most researched group of compounds in seafoods with over 5000 scientific publications on their health-promoting effects (Bahri et al., 2002). ω-3 PUFAs have been cited as beneficial in alleviating, preventing, or reducing the incidence/ severity of a number of diseases including: cardiovascular disease (Burr et al., 1989; Ness et al., 2002), high blood pressure (Morris et al., 1993); blood clotting (Murphy et al., 1999); cancer (Wolk et al., 2006; English et al., 2004), and improved cognitive function (Kalmijn et al., 2004). The ω-3/6 ratio of modern western man is circa 20-fold lower than that of primitive man and indicates massive changes in eating patterns (Block & Pearson, 2006). A second factor is the higher content of ω-6 acids in seafood, and especially farmed fish, which is due to their lower marine and higher vegetable oil diet (see section 7). Castro-Gonzalez et al. (2007) analysed 18 Mexican marine fish for their ω-3 PUFA content in relation to their use as an inherently functional food. Twenty seven percent of the fish showed low (4 to 40), 66% intermediate (7 to 170) and 7% (two species; picuda and sargo) high values (200-300 mg/100 g) for ω-3 PUFAs. These two species are recommended as inherently functional fish. The beneficial effects of ω-3-PUFAs from fish and fish oil on human health is derived from their role in modulating membrane lipid composition and affecting metabolic and signal-transduction pathways (Huang et al., 2009). These authors working with rats demonstrated that EPA and DHA from tuna and salmon oils can be effectively incorporated into tissue membranes.

Other food compounds popular in functional food that may help fight off inflammation are carotenoids. Because carotenoids are very soluble in fat and very insoluble in water, they circulate in lipoproteins, along with cholesterol and other fats. Evidence that LDL oxidation plays a role in the development of atherosclerosis led scientists to investigate the role of antioxidant compounds like carotenoids in the prevention of cardiovascular disease (Kritchevsky, 1999). The thickness of the inner layers of the carotid arteries can be measured noninvasively using ultrasound technology. This measurement of carotid intima-media thickness is considered a reliable marker of atherosclerosis (Bots and Grobbee, 2002). A number of case-control and cross-sectional studies have found higher blood concentrations of carotenoids to be associated with significantly lower measures of carotid artery intima-media thickness (Iribarren et al., 19997; D'Odorico et al., 2000; Rissanen et al., 2000; McQuillan et al., 2001; Rissanen et al., 2003; Dawyer et al., 2004). Additionally, higher plasma carotenoids at baseline have been associated with significant reductions in risk of cardiovascular disease in some prospective cohort studies (Street et al., 1994; Sesso et al., 2004; Rissanen et al., 2001; Ito et al., 2006; Buijsee et al., 2008) but not in others (Evans et al., 1998; Hak et al., 2003; Sesso et al., 2005).

More recently, an analysis of data from the 2003-2006 US National Health And Nutrition Examination Survey (NHANES) in a sample of 2,856 adults found that serum total carotenoid concentration was inversely associated to blood concentrations of two cardiovascular risk factors, C-reactive protein (CRP) and total homocysteine (Wang et al., 2014). HDL-cholesterol concentration was found to be positively associated with α-carotene, β-cryptoxanthin, and lutein/zeaxanthin concentrations, but only the latter was inversely associated with LDL-cholesterol (Wang et al, 2014). Finally, a recent meta-analysis of observational studies reported lower risks of coronary heart disease (-12%) and stroke (-18%) in individuals in the highest versus lowest tertile of blood lutein concentration (Leermakers et al., 2016).

While the results of several prospective studies indicate that people with higher intakes of carotenoid-rich fruit and vegetables are at lower risk of cardiovascular disease (Rimm et al., 1993; Gaziano et al., 1995; Sahyoun et al., 1996; Osganian et al., 2003), it is not yet clear whether this effect is a result of carotenoids or other factors associated with diets high in carotenoid-rich fruit and vegetables.

β-carotene as a dietary supplement

In contrast to the results of observational studies suggesting that high dietary intakes of carotenoid-rich fruit and vegetables may decrease cardiovascular disease risk, four randomized controlled trials found no evidence that β-carotene supplements in doses ranging from 20 to 50 mg/day were effective in preventing cardiovascular disease (Greenberg et al., 1996; Hennekens et al., 1996; Omenn et al., 1996). Based on the results of these randomized controlled trials, the US Preventive Health Services Task Force found good evidence to suggest that β-carotene supplements provided no benefit in the prevention of cardiovascular disease in healthy adults (Moyer, 2014). Thus, although diets rich in β-carotene have generally been associated with reduced cardiovascular disease risk in observational studies, there is no evidence that β-carotene supplementation reduces cardiovascular disease risk (Voutilainen et al., 2006).

Phytoestrogens are proven to improve heart health, specifically in postmenopausal women. They can be used to treat arteriosclerosis, a disease characterized by fatty buildup within the arteries, and seem to do so by regulating many different hormone and chemical levels within the body (Sobenin et al., 2016). Health benefits attributed to a diet rich in phytoestrogens include relief from menopausal symptoms and lowered risk of osteoporosis, heart disease and breast cancer. Phytoestrogens exert their beneficial effects through several mechanisms that slow cell growth and prevent inflammation. In addition, when acting on estrogen receptors, phytoestrogens behave differently from estrogen and more like Selective Estrogen Receptor Modulators (SERMs). An example of a well known SERM is the breast cancer treatment drug tamoxifen, which stimulates estrogen receptors in the uterus and bone, but suppresses them in the breast.

Data from studies that have attempted to show that phytoestrogens are the cause of the lower incidence of fracture and heart disease are inconsistent. One study showed that ingestion of isoflavone-rich soy milk for two years increases lumbar bone density by 2.4%, while another study found no improvement in bone density. Similarly, the phytoestrogen effect on risk of breast cancer remains unknown. It is likely that the effect varies with exposure during different stages of human development. While eating foods rich in phytoestrogen has helped menopausal women in reducing hot flashes and vaginal dryness, data is insufficient and inconsistent to recommend a dose or duration of phytoestrogen intake. Phytoestrogen levels and effects in the body are contingent upon individual intake, absorption, metabolism and time of initiation, which can explain the variation in response and benefit between individuals. Such metabolic differences might explain the apparent inconsistencies in the health effects of phytoestrogens.

Phytoestrogens also occur in medicinal herbs and are widely available as supplements. The dose in supplements is often far higher than those in dietary sources. In terms of safety and effectiveness, supplements should be thought of differently from a diet rich in phytoestrogens.

Blood lipids

Dyslipidemia is the elevation of plasma cholesterol, triglycerides (TGs), or both, or a low high-density lipoprotein level that contributes to the development of atherosclerosis. Causes may be primary (genetic) or secondary (lifestyle). Diagnosis is carried out by measuring plasma levels of total cholesterol, TGs, and individual lipoproteins. Treatment involves dietary changes, exercise, and lipid-lowering drugs.

There is no natural cutoff between normal and abnormal lipid levels because lipid measurements are continuous. A linear relation probably exists between lipid levels and cardiovascular risk, so many people with "normal" cholesterol levels benefit from achieving still lower levels. Consequently, there are no numeric definitions of dyslipidemia; the term is applied to lipid levels for which treatment has proven beneficial. Proof of benefit is strongest for lowering elevated LDL levels. In the overall population, evidence is less strong for a benefit from lowering elevated TG and increasing low HDL levels.

Similarly, HDL levels do not always predict cardiovascular risk. For example, high HDL levels caused by some genetic disorders may not protect against cardiovascular disorders, and low HDL levels caused by some genetic disorders may not increase the risk of cardiovascular disorders. Although HDL levels predict cardiovascular risk in the overall population, the increased risk may be caused by other factors, such as accompanying lipid and metabolic abnormalities, rather than the HDL level itself.

Because of their antioxidant and anti-inflammatory behaviors, flavonoids are associated with cardiovascular disease prevention. Flavonoids may lower the risk of atherosclerosis through protecting LDL cholesterol from free radical damage. They may also improve the quality of blood vessel walls. Several studies have found an association between higher flavonoid intake levels and lowered cardiovascular disease risk across various groups, including postmenopausal women, male smokers and middle-age men and women. For example, a study of more than 10,000 men and women published in 2002 in the American Journal of Clinical Nutrition (Knekt et al., 2002) found that those with higher levels of quercetin had lower rates of ischemic heart disease and those with higher levels of kaempferol, naringenin and hesperetin had lower cerebrovascular disease rates.

Various flavonoids, including quercetin, have shown to be effective at preventing platelet aggregation, according to the Linus Pauling Institute. Platelet aggregation is a known component in heart disease because it contributes to forming blood clots that can lead to strokes and other problems.

The cholesterol lowering effect of plant sterols is well documented in the literature. It is now accepted, after much earlier scientific debate and study, that 4-desmethyl plant sterols or stanols, either in their free or esterified form, decrease blood levels of total cholesterol and LDL-cholesterol through reduction of cholesterol absorption. Generally speaking, properly solubilized free sterols and esterified sterols possess similar cholesterol lowering ability (Hendriks et al., 1999 and Sierksma et al., 1999). In some studies, such comparisons have been flawed because the free sterols were not properly solubilized (Denke 19995). Ostlund et al. (1999) showed that emulsions of sitostanol, mixed with lecithin containing 0.7 g of sterol, reduced cholesterol absorption considerably, whereas less effect was seen with sitosterol in crystalline form.

Important issues that remain to be verified regarding the cholesterol lowering effect of plant sterols include (Berger et al., 2004):

- efficacy of low dose of plant sterols
- the effect diet background on plant sterol efficacy
- the efficacy of plant sterols when incorporated in food other than fat spread
- the optimal number of plant sterols servings
- the relative efficacy of plant sterols among different populations

Hypertension

Elevated blood pressure is a significant risk factor for CVD. Moderately elevated blood pressure is a very attractive target for functional food product development, as it is one of a relatively small group of health conditions where the consumer can readily assess the effect of alternative therapies (Muir, 2005). Fish in common with other foods contain anti-hypertensive peptides known as angiotensin I-converting enzyme (ACE) inhibitors (Gormley, 2006). Many of the active compounds are contained in hydrolysates from fish waste. ACE inhibitory peptides lower blood pressure by limiting the vasoconstrictory effects of Angiotensin II and potentiating the vasodilatory effects of bradykinin (De Leo et al., 2009). Matsufuji et al. (1994) isolated 13 ACE inhibitors from sardine mussel and most were competitive inhibitors of ACE with the potential to lower blood pressure. Vercruysse et al. (2005) reviewed the topic of ACE inhibitors derived from enzymatic hydrolysates of fish muscle and concluded that these peptides offer great potential as antihypertensive agents in functional foods.

Enzyme-treated fish protein hydrolysates (FPHs) can also be used as cardio-protective (antiatherogenic) agents in the form of a functional food or pharmaceutical (Berge, 2005).

Table 20: Bioactive compounds in food for cardiovascular care

Adapted from: Mohammed, 2014

Category	Examples	Structures of bioactive compounds	
Oils and fats	Fish oil olive oil, flax seed oil, OMEGA 3 fatty acid, Omega 6 fatty acid	Dodecahexaenoic acid 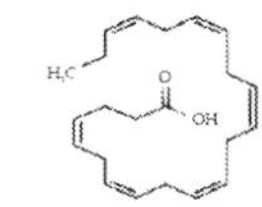	Oleic acid 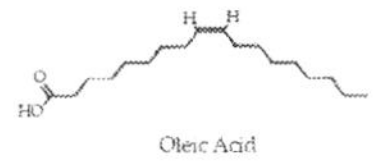
Digestive Enzymes Inhibitors (Alpha-glucosidase, alpha-amylase & Gastrointestinal lipase inhibitors	Green and black tea, berries (lingonberry, bearberry,arctic bramble, cloudberry, strawberry, raspbery, blueberry, grarden pea)	Epigallocatechin-3-gallate, 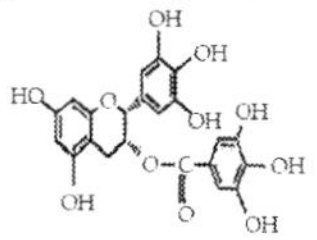	Tannins 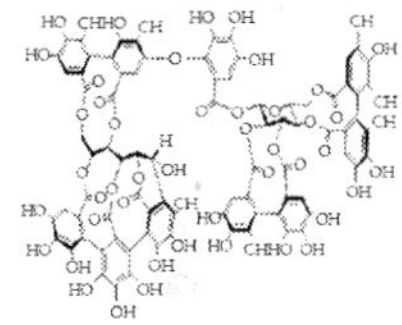
Beverages	Black and green Tea, cacoa coffee, herbal tea	chlorogenic acid (5-caffeoyl quinic acid 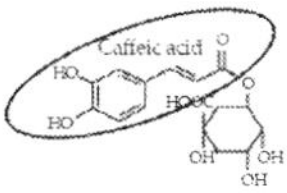	Caffeine 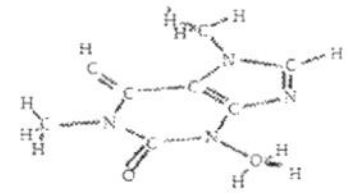
Fruits	Berries, apples, citrus, grapes, pomegranates, plums, rose hips, rosele, watermelon, mangoes	Resveratrol 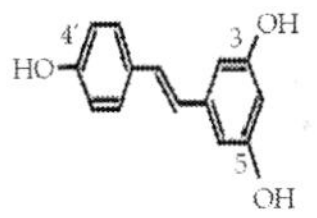	Anthocyanin 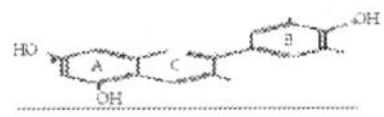
Vegetables	Onions and gralic, brassica (cabbage family), tomatoes, potatoes, bitter melon, mushrooms, seaweeds	Alliin, all-cin 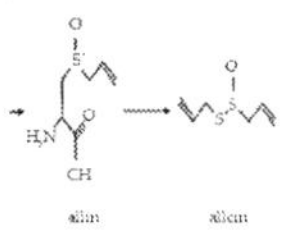	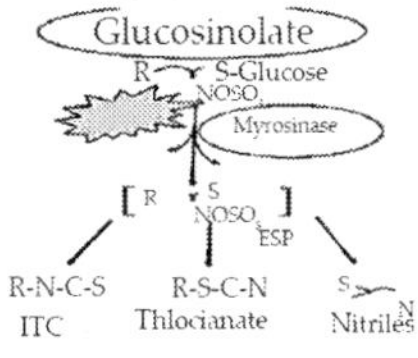
Grains and legumes	Nuts, sesame, whole grain cereals, soya,	beta-glucan 	Isoflavones 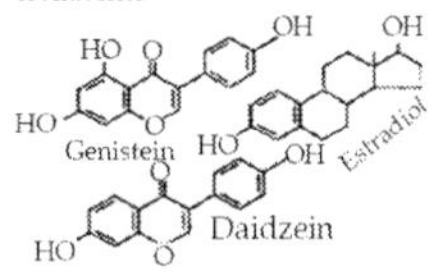

Neurological Benefits

Neurodevelopment relates to the growth and development of the brain and central nervous system and includes brain function, emotion, learning ability, memory and long-term cognition. The most widely studied functional food ingredients related to brain health are essential fatty acids.

DHA is essential for featal development with expectant mothers haveing an increased need for DHA which they can cover from a combination of supplementation (e.g. capsules) and by eating oily fish (Nys and Debruyne, 2011). For pre-term infants DHA supplementation has a beneficial effect early in life on cognitive development at greater than 12 months of age. However, for term infants or for healthy children older than two there is no evidence for beneficial effects on cognitive performance following DHA supplementation (Eilander et al., 2007). However, some inconsistencies between studies may be due to genetic effects which are not yet fully understood. For example, some breast-fed children attain higher IQ scores than those not fed breast milk, presumably because of PUFAs uniquely available in breast milk. However, in those with a quite common genetic variant (FADS2) involved in fat metabolism, the beneficial effect of breast feeding was not observed (Caspi et al., 2007).

Current evidence suggests a protective effect of ω-3 PUFAs against dementia and over 100 clinical trials are running globally on 7 cognitive health and nutrition. A study in the Nehru Science Centre, Mumbai, has shown that DHA decreases progression of neurodegenerative disorders in older age. In contrast, a recent UK study on young adults (18-29 years) showed that DHA supplementation increased blood flow to the brain but was not accompanied by consistent improved cognitive performance in solving computerised cognitive tasks relative to a control group (Jackson et al., 2012). The so-called EPOCH trial is investigating the effect of ω-3 PUFAs on cognitive ageing and wellbeing in cognitively healthy older adults (391 individuals; 54% females). It includes persons with an alternative form of a gene which is thought to modify the effect of DHA supplementation (Danthiir et al., 2011).

Tan et al. in 2012 confirmed that an EPA/DHA rich diet protects against brain ageing. Over 1500 dementia free patients (average age 67) underwent brain scans and were tested for mental function, body mass, and EPA/DHA content of their red blood cells. The group was divided in four; the bottom quartile having the lowest EPA/DHA levels in their blood and the top quartile the highest, with the two middle quartiles having intermediate levels. Those in the bottom quartile had lower brain volumes than those in the other three quartiles and also scored lower on visual memory tests and executive function, such as problem solving and multitasking. A further study by Virtanen et al. (2013) suggests that the level of circulating ω-3 PUFAs (a biomarker of regular fish consumption) may be beneficial for the prevention of certain subclinical brain abnormalities that are commonly observed in the elderly.

Cognitive decline

The brain, as does the rest of the human body, changes with the aging process. Many people notice gradually increasing forgetfulness as they age. It may take longer to think of a word or to recall a person's name. But consistent or increasing concern about overall mental performance may suggest mild cognitive impairment (MCI).MCI is an intermediate stage between the expected cognitive decline of normal aging and the more-serious decline of dementia. It can involve problems with memory, language, thinking and judgment that are greater than normal age-related changes.

Observational studies have suggested that dietary lutein may be of benefit in maintaining cognitive health (Kang et al, 2005; Morris et al., 2006). Among the carotenoids, lutein and its isomer zeaxanthin are the only two that cross the blood-retina barrier to form macular pigment in the eye. Lutein also preferentially accumulates in the brain (Johnson et al., 2013; Vishwanathan et al., 2014). Recent studies suggested that lutein and zeaxanthin concentrations in the macula were correlated with brain lutein and zeaxanthin status and might be used as a biomarker to assess cognitive health (Vishwanathan et al., 2014; Feney et al., 2013; Renzi et al., 2014). Additionally, in the Georgia Centenarian Study, the analysis of cross-sectional data from 47 centenarian decedents showed a positive association between post-mortem measures of brain lutein concentrations and pre-mortem measures of cognitive function (Johnson et al., 2013). Brain lutein concentrations were found to be significantly lower in individuals with mild cognitive impairment compared to those with normal cognitive function (Johnson et al., 2013). In a small, four-month, double-blind, placebo-controlled study in older women (ages, 60 to 80 years) without cognitive impairment, supplementation with lutein (12 mg/day) and zeaxanthin (~0.5 mg/day) significantly improved cognitive test performance (Johnson et al., 2008). Nevertheless, the Age-related Eye Disease Study 2 (AREDS2) failed to show an effect of supplemental lutein (10 mg/day) and zeaxanthin (2 mg/day; supplementation for a median of 4.7 years) on the cognitive test scores of 3,741 participants (mean age, 72.7 years) (Chew et al., 2015).

Flavonoids' anti-inflammatory and antioxidant effects may help protect against neurodegenerative diseases like Alzheimer's and Parkinson's. In animal studies, flavonoid levels have been positively correlated with reduced risk of these diseases, but human studies have yielded inconclusive results. For example, a large-scale study published in 2000 in the European Journal of Epidemiology (Commenges et al., 2000) found that among elderly men and women, those with the highest levels of flavonoids had a 50 percent lower risk of developing dementia over the next five years than those with the lowest levels of flavonoid intake. On the other hand, a study published in 2002 in JAMA (Engelhart et al., 2002) found that among men, the only group that saw a reduced risk of Alzheimer's when increasing their flavonoid intake was smokers. The same results were seen in terms of Parkinson's disease, according to a study published in 1997 in Archives of Neurology (de Rijk et al., 1997).

Flavonoids may also increase blood flow to the brain, improving cognitive function. A study published in 2007 in the American Journal of Epidemiology (Letenneur et al., 2007) found that elderly men and women with higher flavonoid intake had better cognitive performance at the start of the study and significantly less age-related cognitive decline over the next 10 years than those with lower flavonoid intake.

Gastrointestinal Benefits

Gastrointestinal diseases are those that affect any section of the gastrointestinal tract, from the oesophagus to the rectum, and the accessory digestive organs including the liver, gall bladder and pancreas. The term encompasses acute, chronic, recurrent or functional disorders and covers a wide range of diseases, including inflammatory bowel disease and functional dyspepsia.

Irritable Bowel Syndrome

Irritable Bowel Syndrome (IBS) is a common condition characterised by pain, distension and altered bowel habit. Evidence suggests functional foods containing probiotics improve gastrointestinal transit, however, data are limited by short follow-up periods and evaluation in selected populations. Various systematic reviewshave evaluated the impact of probiotics on IBS. These reviews have highlighted the considerable heterogeneity in study findings and also in outcomes, probiotic combinations and study design. Therefore, despite the number of studies, the evidence base remains poor with data coming largely from small or uncontrolled trials. Overall, within the published literature and systematic reviews there is a suggestion that probiotics confer benefit although limitations of available evidence and possibility of publication bias are acknowledged. Larger well-designed studies of probiotics in IBS are therefore required.

The pathophysiology of IBS is poorly understood and is currently thought to represent a complex interplay among the gut microbiota, low-grade inflammation, impaired mucosal barrier function, visceral hypersensitivity, gut motility, and alterations in the gut-brain axis. In any individual patient, 1 or more of these factors may interact to generate symptoms. Although up to 50% of patients report postprandial exacerbation of symptoms, few studies have critically assessed the role of diet in IBS. Furthermore, although many patients with IBS adopt any one of a host of dietary changes in an attempt to alleviate their symptoms, there has been, up until recently little scientific basis for any dietary recommendation in IBS.

In a double-blind randomized controlled trial (Roberts et al., 2013) of a commercially available probiotic fermented dairy product (yoghurt) containing bifidobacterium animalis DN-173010 in community based IBS patients with a constipation element, both the inactive (control) and active product groups demonstrated significant improvement across a range of symptom and quality of life outcomes. Improvement was maintained throughout the 12-week period of observation. Reductions in symptoms were, however, reported at similar levels in both groups when assessed in weeks 4 and 8. At week 12 there was the suggestion of a benefit of the control product over the active formulation but this is likely to be attributable to differential drop-outs between groups. The sustained and large improvement observed in both groups suggests there may be benefit from regular consumption of a dairy product but does not suggest any additional benefit of the addition of a probiotic to such products. If such a benefit is accrued the mechanism through which this works is unclear and product effect, regularisation of eating habits and increased fluid intake are all possible explanations. The fact that benefits were also accrued by control participants supports these potential mechanisms of effect. In the absence of a non-intervention group this effect cannot be confirmed but the substantial effect size is worthy of further consideration and exploration of sub-groups in which greatest benefit is accrued would be advisable. In this study there was the suggestion that drop-out may be related to baseline symptom severity with those remaining in the study having greater baseline symptom severity scores than those withdrawing (IBS SSS 264.15 versus 238.09, $p = 0.04$). This could suggest greater benefit was accrued by those with more severe symptoms encouraging continued participation or may reflect and greater motivation to pursue therapy in this group.

Research is beginning to support a potential role for probiotics in IBS, with variability depending on the use of different probiotic strains, their ability to adhere to and colonize in the GI tract, and the number of colony-forming units an individual ingests.

According to a 2008 article on prebiotics and probiotics in dietetic practice published in the Journal of the American Dietetic Association, Bifidobacterium infantis 35264 in particular has been shown in published human studies to be effective in treating IBS. Catsos recommends a probiotic based on bifidum bacteria, such as B infantis 35264, because bifidobacterium is the only group of common gut bacteria that does not produce any gases. Many IBS nutrition experts find prebiotics and probiotics to be helpful for their clients but stress that they should not be used as a first-line defense.

Digestion

It is often stated that the human GI tract serves 2 main purposes: acting as a barrier to the external environment and as the main portal of entry for nutrients. The duality of this paradigm, however, may obscure an important feature of some nutrients, namely, to preserve the integrity and function of the GI mucosa itself. Because GI epithelial cells are the primary interface between ingested nutrients and the blood and lymph streams, it is not surprising that these cells are dependent on both luminal and bloodstream sources for their nutrition. Other cells of the GI tract, including M cells, enteroendocrine cells, intraepithelial lymphocytes, and the multiple cell types of the lamina propria are all affected by intra- and extraluminal nutrient intake.

Most of the clinical studies supporting the use of probiotics in the prevention or treatment of GI disease were carried out in pediatric patients (Pearce and Hamilton, 1974), although increasing data are being reported from adult trials. Early studies that used a commercial preparation of dried L. acidophilus and Lactobacillus bulgaricus (Lactinex Lactobacillus; Becton Dickinson Consumer Products, Franklin Lakes, NJ) showed no significant effects in the prevention of traveler's diarrhea (de dios Pozo-Olano et al., 1978) or diarrhea due to enterotoxigenic E. coli (Clements et al., 1981). More recent preparations, especially Lactobacillus casei strain GG (Oksanen et al., 1990 and Siitonen et al., 1990) and combinations of probiotics, however, have proven more efficacious.

Many human studies confirm that fructose oligosaccharides have a bifidogenic effect on human colonic endogenous flora (Gibson and Roberfroid, 19995; Sghir et al., 1998; Gibson and Fuller, 2000). Several clinical studies also showed a bifidogenic effect of inulin-type fructans in humans (Bouhnik et al., 1997; Roberfroid, 1998; Van Loo et al., 1999; Alles et al., 1999; Ito et al., 1993). In these studies, the growth of a bacterium, in this case bifidobacteria, was measured in the stool of human volunteers fed varying amounts of prebiotics. However, when stool slurries were incubated with a specific prebiotic and changes in bifidobacteria were quantitated, the results were mixed and a dose-response curve could not be obtained (Roberfroid, 1998). These observations suggest that the bifidogenic effect is not simply that prebiotics are preferential substrates for bifidobacteria but that prebiotics interact with other bacteria and are associated with environmental changes, such as changes in luminal pH, and other factors to achieve the overall bifidogenic effect. No clinical studies have been done to determine how long the bifidogenic effect persists after supplementation and whether the effect is sustained after the prebiotic is discontinued.

These studies generally were small trials among healthy adult subjects, and there have been fewer studies among larger numbers of subjects. A study among young children attending daycare showed that prebiotic-supplemented cereal was associated with fewer episodes of diarrhea with fever than was control cereal, although the overall rate of diarrhea episodes was not different between the groups (Saavedra et al., 1999). A recent study in which Peruvian children were similarly supplemented did not show a reduction in diarrhea or other infections (Duggan et al., 2001).

With regard to other functions of protective nutrients and functional foods in man, some studies suggest that prebiotics improve Ca^{2+} absorption from the human colon (van den Heuvel et al., 1999 and Coudray et al., 1997). Other studies, however, had equivocal results (van Vliet, 19997). No definitive clinical studies support the animal studies to suggest that other minerals (Mg^{2+}, Fe^{2+}, and Zn^{2+}) are similarly affected.

Several clinical studies were conducted to determine the effect of prebiotics on lipid metabolism in humans. In a randomized, double-blind trial among adults with moderate hyperlipidemia, prebiotic supplementation was associated with reduction in fasting insulin and triacylglycerol concentrations. These effects did not persist after the supplement was stopped (Jackson et al., 1999). Other studies showed no effect of prebiotics on lipids (Alles et al., 1999). These equivocal results were particularly true when healthy adults with normal lipid concentrations were studied (van Vliet, 19997).

As promising as prebiotics may be as protective nutrients, many clinical studies must be done before prebiotics can be recommended as a food additive (infant formula, yogurt, etc) or as a dietary supplement. (Note that prebiotics have already been introduced as food additives in Europe and Japan.) Studies to determine the longevity of the prebiotic effect are needed before these fructose oligosaccharides can be considered as a protective nutrient (Commings et al., 2001). In addition, the possible effect of combining prebiotics with probiotics should be evaluated. No human studies have been conducted to confirm the suggested in vitro animal study effect of prebiotics on carcinogenesis. Long-term trials with prebiotics, perhaps first among colon cancer–prone patients, would be required.

Cancer Preventing Benefits

Cancers are diseases in which abnormal cells divide without control and can invade nearby tissues. Cancer cells can also spread to other parts of the body through the blood and lymph systems. There are several main types of cancer. Carcinoma is a cancer that begins in the skin or in tissues that line or cover internal organs. Sarcoma is a cancer that begins in bone, cartilage, fat, muscle, blood vessels, or other connective or supportive tissue. Leukemia is a cancer that starts in blood-forming tissue, such as the bone marrow, and causes large numbers of abnormal blood cells to be produced and enter the blood. Lymphoma and multiple myeloma are cancers that begin in the cells of the immune system. Central nervous system cancers are cancers that begin in the tissues of the brain and spinal cord.

Lung cancer

Several large prospective cohort studies, including the Nurses' Health Study (NHS) and the Health Professionals Follow-up Study (HPFS), have examined potential associations between carotenoid intake and/or blood concentrations and lung cancer (Gallicchio et al., 2008). In a meta-analysis of eight prospective cohort studies, including NHS and HPFS, the highest versus lowest quantile of total carotenoid intake was significantly associated with a 21% reduced risk of lung cancer. For the individual carotenoids, the risk of lung cancer was estimated to be 20% and 14% lower with the highest versus lowest intakes of β-cryptoxanthin and lycopene, respectively. In contrast, dietary intakes of β-carotene, α-carotene, and lutein/zeaxanthin were not found to be significantly linked to a reduced risk of developing lung cancer (Gallicchio et al., 2008). In addition, an analysis of the pooled results of 11 nested case-control and four prospective cohort studies found no association between serum concentrations of total carotenoids, β-carotene, α-lycopene, β-cryptoxanthin, and lutein/zeaxanthin and lung cancer. Only high versus low serum lycopene concentrations could be linked to a 29% lower risk of lung cancer (Gallicchio et al., 2008). Any protective effect of dietary carotenoids against the development of lung cancer is likely small and not statistically significant (Gallicchio et al., 2008).

In Finland, the A-Tocopherol Beta-Carotene (ATBC, 1994) cancer prevention trial evaluated the effects of 20 mg/day of β-carotene and/or

50 mg/day of α-tocopherol on more than 29,000 male smokers, and in the United States, the β-Carotene And Retinol Efficacy Trial (CARET) evaluated the effects of a combination of 30 mg/day of β-carotene and 25,000 IU/day of retinol (preformed vitamin A) in 18,314 men and women who were smokers, former smokers, or had a history of occupational asbestos exposure (Omenn et al., 1996). Unexpectedly, the risk of lung cancer in the groups taking β-carotene supplements was increased by 16% after six years in the ATBC participants and by 28% after four years in the CARET participants. In contrast, the Physicians' Health Study (PHS) examined the effect of β-carotene supplementation (50 mg every other day) on cancer risk in 22,071 male physicians in the United States, of whom only 11% were current smokers (Hennekens et al., 1996). In that lower risk population, β-carotene supplementation for more than 12 years was not associated with an increased risk of lung cancer. In addition, in the Linxian General Population trial conducted in about 29,000 Chinese participants, randomization to 15 mg/day of β-carotene, 30 mg/day of α-tocopherol, and 15 μg/day of selenium, was not found to be associated with lung cancer mortality 10 years after the intervention ended (Kamangar et al., 2006).

Although the reasons for the increase in lung cancer risk are not yet clear and several mechanisms have been proposed (Veeramachaneni and Wang, 2009). The US Preventive Services Task Force estimated that the risks of high-dose β-carotene supplementation outweigh any potential benefits for cancer prevention and recommended against supplementation, especially in smokers or other high-risk populations (Moyer, 2014).

Isothiocyanates combat carcinogens by neutralizing them, reducing their poisonous effect and stimulating the secretion of carcinogens of carcinogens. Isothiocyanates act by inhibition of cell proliferation and induction of apoptosis. The isothiocyanates with the stongest anticancer effects are phenylethylisothiocyanate, benzylisothiocyanate and 3-phenylpropylisothiocyanate. Studies have shown that isothiocyanates help to prevent lung cancer and esophageal cancer. Isothiocyanates can also lower the risk of other cancers, including gastrointestinal cancer.

Prostate cancer

Prostate cancer is the most common cancer affecting US men, more prevalent than lung and colorectal cancers (CDC, 2015). Several early prospective cohort studies have suggested that lycopene-rich diets were associated with significant reductions in the risk of prostate cancer, particularly more aggressive forms (Giovannucci, 2002). However, pooled data analyses of observational studies that examined potential links between dietary intakes and/or circulating concentrations of lycopene and risk of prostate cancer have given mixed results. An early meta-analysis that combined the results of 10 case-control and four prospective studies found that men with the highest intakes of raw tomatoes, cooked tomatoes, or dietary lycopene had a 11 to 19% lower risk of prostate cancer (Etminan et al., 2004). In addition, pooled data from two case-control and five nested case-control studies showed a 26% lower risk of prostate cancer in participants with the highest serum concentrations of lycopene (Etminan et al., 2004). Most recently, a meta-analysis of observational studies found no association of prostate cancer risk with dietary lycopene intakes (10 case-cohort and two prospective cohort studies) but an inverse association with blood lycopene concentrations (two case-control, nine nested case-control, and one cohort studies) (Wang et al., 2015 and 2015b). Additionally, a meta-analysis of 15 nested case-control studies conducted by the Endogenous Hormones, Nutritional Biomarkers, and Prostate Cancer Collaborative Group showed an inverse association between circulating lycopene concentrations and risk of advanced stage and/or aggressive prostate cancer, while no association was found with risk of non-aggressive or localized disease (Key et al., 2015).

While there is considerable scientific interest in the potential for lycopene to help prevent prostate cancer, it is not yet clear whether the prostate cancer risk reduction observed in some observational studies is related to lycopene itself, other compounds in tomatoes, or other factors associated with lycopene-rich diets (Melendez-Martinez et al., 2015). Of note, the 2014 World Cancer Research Fund International report on Diet, Nutrition, Physical Activity, and Prostate Cancer suggested the need for better designed studies to establish whether consumption of lycopene-containing foods could be linked to a lower risk of prostate cancer (WCRF, 2014).

To date, a few short-term, dietary intervention studies using lycopene in patients with precancerous prostate lesions (high grade prostatic intraepithelial neoplasia; HGPIN) or prostate cancer have been completed. Specifically, two small randomized controlled studies have recently examined the effect of lycopene supplementation for up to six months in men with HGPIN (Gann et al., 2015; Gontero et al., 2015). The consumption of 30 to 35 mg/day of supplemental lycopene in the form of tomato extract (Gann et al., 2015), or together with selenium (55 mg/day) and green tea catechins (600 mg/day) (Gontero et al., 2015), showed no benefit on the rate of progression to prostate cancer at six-month (Gann et al., 2015; Gontero et al., 2015) and 37-month follow-ups (Gontero et al., 2015). Additionally, a randomized controlled trial in men with localized prostate cancer found that supplementation with 15, 30, or 45 mg of lycopene (until prostatectomy) did not significantly increase plasma lycopene concentration, modify the ratio of steroid hormones in blood, or reduce the concentration of markers of proliferation (i.e., prostate-specific antigen [PSA] and Ki-67) compared to placebo (Kumar et al., 2008). In another trial, 54 patients with metastatic prostate cancer were randomized to orchidectomy alone or orchidectomy plus 4 mg/day of lycopene (Ansari and Goupta, 2003). The proportion of complete clinical response to treatment — assessed by serum PSA and/or bone scan returning to normal — and patient survival rate were found to be significantly higher in patients supplemented with lycopene (Ansari and Goupta, 2003). Large-scale, controlled clinical trials are needed to further examine the safety and efficacy of long-term use of lycopene supplements for prostate cancer prevention or treatment.

Other types of cancer

In a meta-analysis of 12 prospective cohort studies, no association was found between total and individual carotenoid intake and risk of breast cancer, except with β-carotene for which a 5% reduction in breast cancer risk was estimated for every 5 mg/day increment in consumption (Aune et al, 2012). In a pooled analysis of 14 nested case-control studies and one follow-up study of a clinical trial, reductions in breast cancer risk were found to be associated with blood concentrations of total carotenoids (-26%), α-carotene (-20%), and lutein (-30%) (Aune et al., 2012). Another study that recalibrated data for consistency across eight large prospective cohorts before pooled analysis found reduced breast cancer risk to be associated with the highest versus lowest quintile of blood concentrations of total carotenoids (-21%), β-carotene (-17%), and lycopene (-22%) (Eliassen et al., 2012). Further analyses found an inverse association between the blood concentrations of β-carotene and α-carotene and risk of estrogen receptor-negative (ER-), but not estrogen receptor-positive (ER+), breast tumors (Eliassen et al., 2012). A similar result was recently reported in a case-control study nested within the multicenter, large, European Prospective Investigation into Cancer and Nutrition (EPIC) study (Bakker et al., 2016).

Another case-control study nested within the EPIC study found a 31% lower risk of colorectal cancer with the highest versus lowest quartile of β-carotene intake, while no associations were shown with blood concentrations of other carotenoids or total carotenoids (Leenders et al., 2014). The nested case-control study was included in a recent meta-analysis of 22 observational studies that failed to find associations between carotenoid intakes and colorectal cancer (Panic et al., 2016).

Pooled data from mostly case-control studies also suggested that higher intakes of individual carotenoids, especially β-cryptoxanthin and lycopene, might be associated with a reduced risk of cancers of the mouth, pharynx, and larynx (Leoncini et al., 2015). However, the results of case-control studies are more likely to be distorted by bias than results of prospective studies. Examining blood concentrations of each carotenoid in relation to cancer subsites may help overcome limitations associated with dietary data and differences in carotenoid absorption.

Cancers related to hormone production may be managed in part by adjusting the hormone levels in the body by eating the right foods. Phytoestrogens have been studied extensively in connection with breast and ovarian cancers, with many positive results showing they can actually be natural cancer treatments for some. A 2009 study of over 5,000 women diagnosed with breast cancer showed a significant decrease in death and recurrence of the disease of the patients on a diet rich in non-soy phytoestrogens, a finding that echoed a 1997 questionnaire study of breast cancer patients (Shu et al., 2009 and Ingram et al., 1997). Regarding breast cancer in particular, it seems that apigenin might be the best of the phytoestrogens in reducing breast cancer cell growth (Scherbakov and Andreeva, 2015).

The jury is still out on exactly how and when phytoestrogens are most efficient in fighting hormonal cancers. Depending on the state of menopause, individual body makeup and the times at which high levels of soy are part of one's diet, phytoestrogens may or may not be beneficial for cancer prevention and/or treatment. (Kyro et al., 2015).

Today we are at a new frontier in nutritional science. The transition from 'adequate' to 'optimal' nutrition. It is here that functional foods will have a pivotal role in reducing diet-related diseases.

Pro-aging Benefits

Pro-aging medicine utilizes and prescribes medications when necessary, reintroduces whole food nutrition, appropriate supplementation, and natural hormonal replacement therapy (HRT). Hormonal blood levels are ordered, as well as other markers useful for prevention and reversal of declining health. This type of medicine has been coined internal plastic surgery and may greatly benefit from the functional food market.

Weight management

The first indication of the role of the microbiota in the pathogenesis of obesity was published several years ago.With the same diet intake, conventionally raised mice gained more body weight and body fat than germ free mice. Furthermore, transplantation of microbiota harvested from conventionally raised mice into germ free mice resulted in an increase in body weight and a decrease in insulin sensitivity (Backhed et al., 2004). Not only were germ-free mice leaner than conventionally raised mice, but they were also resistant to western-type high-fat diet induced obesity (Backhed et al., 2007).

Obesity is correlated with different bacterial communities and that a gradual transition occurs in humans from the obese microbiota to the lean microbiota during a course of a restrictive energy intake. People whose colon contains a relatively low diversity of bacteria and bacterial genes were found to contain higher levels of body fat and inflammation than those with high gut-microbial richness (Fang and Evans, 2013; Le et al., 2013). The data suggesting a relationship between obesity and the diversity of intestinal microbiota open the future possibility of treating obesity by gut microbiota manipulation. Regarding "obese bacteria", it should be noted, they are also associated with normal weight gain during the third trimester of pregnancy (Koren et al. 2012), where they may be highly beneficial, as they promote energy storage in fat tissue and provide for the growth and development of the foetus and subsequently for the production of mother's milk for the infant after birth, all of which are central to the fitness of mammals. The origins of host-microbial interactions that underlie much of the present-day obesity epidemic may lie in reproductive biology and survival mechanisms correlated with food shortage in the past.

Amongst the different foods that are included in the human diet are those that are nicknamed "superfoods." These include such foods as pomegranate, blueberries, green tea, salmon, broccoli, garlic and goji berries. Through the term "superfood" is considered to be more of a marketing term, professionals admit that these types of foods are nutraceutical-rich and have been shown to be healthy and disease preventing in animal trials (Yi et al. 2005; Malin et al., 2011) as well as some human clinical trials (Aviram et al. 2000; Kris-Etherton and Keen 2002; Hooper et al. 2008). Many of the nutraceuticals contained in the superfoods are broken down in the human gut by the microbiota and this transformation often turns them into physiologically active materials.

Premkumar (2014) noted that flavonoids are also associated with inflammation and weight loss. Flavonoid content can relieve inflammation and decrease the levels of an appetite-suppressing hormone, leptin. Leptin plays an important role in food consumption because mice with mutations in leptin or its receptor become obese, and these animals are used as models for studying diabetes and obesity.

The phytoestrogen genistein is highlighted in a study to determine the effects of phytoestrogens on obesity. Due to its various effects, genistein seems to have the ability to regulate obesity, although the reasons why are unclear (Dang, 2009). This weight loss benefit requires more research to draw specific conclusions, but the findings are encouraging.

Age-related macular degeneration

In Western countries, degeneration of the macula, the center of the eye's retina, is a leading cause of blindness in older adults. Long-term blue light exposure and oxidative damage in the outer segments of photoreceptors may lead to drusen and/or pigment abnormalities in the macula, increasing the risk of age-related macular degeneration (AMD) and central blindness.

The carotenoids found in the retina are lutein and zeaxanthin, which are both of dietary origin, and meso-zeaxanthin, which is derived from lutein. These three carotenoids are present in high concentrations in the macula (known as macular pigment), where they are efficient absorbers of blue light. They may prevent a substantial amount of the blue light entering the eye from reaching the underlying structures involved in vision and protect against light-induced oxidative damage, which is thought to play a role in the pathology of age-related macular degeneration. It is also possible, though not proven, that lutein, zeaxanthin, and meso-zeaxanthin, act directly to neutralize oxidants formed in the retina. Although lutein and zeaxanthin are different compounds, they are both classified as xanthophylls and nonprovitamin A carotenoids. Some methods used to quantify lutein and zeaxanthin in food do not separate the two compounds, so they are typically reported as lutein and zeaxanthin or lutein + zeaxanthin. Both pigments are present in a variety of fruit and vegetables. Dark green leafy vegetables like spinach and kale are particularly rich sources of lutein but poor sources of zeaxanthin (Hendler and Rorvik, 2008). Although relatively low in lutein, egg yolks and avocados are highly bioavailable sources of lutein. Good sources of dietary zeaxanthin include yellow corn, orange pepper, orange juice, honeydew (melon), and mango (Hendler and Rorvik, 2008). Some foods containing lutein and zeaxanthin are listed in Table 21.

Table 21: Lutein + Zeaxanthin content of selected foods

Food	Serving	Lutein + Zeaxanthin (mg)
Spinach, frozen, cooked	1 cup	29.8
Kale, frozen, cooked	1 cup	25.6
Turnip greens, frozen, cooked	1 cup	19.5
Collards, frozen, cooked	1 cup	18.5
Dandelion greens, cooked	1 cup	9.6
Mustard greens, cooked	1 cup	8.3
Summer squash, cooked	1 cup	4.0
Peas, frogen, cooked	1 cup	3.8
Winter squash,baked	1 cup	2.9
Pumpkin, cooked	1 cup	2.5
Brussel sprouts,frogen, cooked	1 cup	2.4
Broccoli,frogen, cooked	1 cup	2.0
Sweet yellow corn, boiled	1 cup	1.5
Avocado, raw	1 medium	0.4
Egg yolk, raw	1 large	0.2

Increasing dietary consumption of lutein and zeaxanthin was shown to raise their serum concentration and macular pigment density (Hammond et al., 1997; Mares et al., 2006). Some, but not all, observational studies have provided evidence that higher intakes of lutein and zeaxanthin are associated with lower risk of age-related macular degeneration (Mares-Perlman, 2002). While cross-sectional and retrospective case-control studies found that higher levels of lutein and zeaxanthin in the diet (Sedon et al., 1994; Snellen et al., 2002; Mares-Perlman et al., 2001), blood (Gale et al., 2003), and retina (Bone et al., 2001; Beatty et al., 2001) were associated with a lower incidence of AMD, several prospective cohort studies found no relationship between baseline dietary intakes or serum concentrations of lutein and zeaxanthin and the risk of developing

AMD over time (Mares-Perlman et al., 1995; Mares-Perlman et al., 1996; Flood et al., 2002; Cho et al., 2004;). A recent report examined the association between the incidence of AMD and calculated dietary intakes and predicted plasma concentrations of lutein and zeaxanthin in older adults (≥50 years) from two large prospective cohorts, the Nurses' Health Study (NHS; 63,443 women) and the Health Professionals Follow-up Study (HPFS; 38,603 men), followed for 26 years and 24 years, respectively (Wu et al., 2015). The highest versus lowest quintile of predicted plasma lutein and zeaxanthin scores was associated with a 41% lower risk of advanced AMD, yet no association was found with intermediate AMD. The evidence also suggested that the consumption of about 6 mg/day of lutein and zeaxanthin from fruit and vegetables (compared with less than 2 mg/day) may decrease the risk of advanced AMD (Wu et al., 2015).

The first randomized controlled trial in patients with atrophic AMD found that supplementation with 10 mg/day of lutein slightly improved visual acuity after one year compared to a placebo (Richer et al., 2004). The effects of long-term lutein supplementation on atrophic AMD were further investigated in combination with antioxidant vitamins and minerals in the Age-Related Eye Disease Study 2 (AREDS2), a multicenter, randomized, double-blind, placebo-controlled trial. In AREDS, oral supplementation with β-carotene (15 mg/day), vitamin C (500 mg/day), vitamin E (400 IU/day), zinc (80 mg/day as zinc oxide), and copper (2 mg/day as cupric oxide) for five years was shown to reduce the risk of developing advanced AMD by 25% (AREDS2, 2013). In the AREDS2 study conducted in 4,203 participants at risk for developing late AMD, supplemental lutein (10 mg/day) and zeaxanthin (2 mg/day), in combination with β-carotene, vitamin C, vitamin E, zinc, and copper (the 'AREDS' formulation), did not slow the progression to advanced AMD, although subgroup analyses revealed a benefit in those with the lowest dietary intakes of lutein + zeaxanthin (AREDS2, 2013). A total of 3,036 subjects were further randomized to various combinations of carotenoids; supplementation with lutein and zeaxanthin significantly reduced the risk of progression to late AMD and to neovascular AMD compared to supplementation with β-carotene (AREDS2, 2014).

Several smaller (n=30-433) randomized controlled trials also suggested that supplementation with xanthophyll carotenoids would be beneficial in the management of AMD (Scripsema et al., 2015). A meta-analysis of eight trials that examined the effect of supplemental lutein (6 to 20 mg/day) or/and zeaxanthin (0 to 10 mg/day) in 1,176 AMD subjects for up to 36 months found improvements in visual acuity and contrast sensitivity with increased levels of xanthophyll carotenoids (Liu et al., 2015). Finally,a recent randomized, placebo-controlled trial found daily consumption of a buttermilk drink with egg yolks enriched with lutein (1.4 mg), zeaxanthin (0.2 mg), and an omega-3 polyunsaturated fatty acid (DHA; 160 mg) for one year improved visual acuity and macular pigment optical density in subjects with drusen and/or retinal pigment abnormalities (about half of them being classified as having early AMD) (van der Made et al., 2016).

The first randomized controlled trial designed to examine the effect of a carotenoid supplement on AMD used β-carotene in combination with vitamin C, vitamin E, and zinc because lutein and zeaxanthin were not commercially available as supplements at the time the trial began (AREDS 2001). Although the combination of antioxidants and zinc lowered the risk of developing advanced macular degeneration in individuals with signs of moderate-to-severe macular degeneration in at least one eye, it is unlikely that the benefit was related to β-carotene since it is not present in the retina. Supplementation of male smokers in Finland with 20 mg/day of β-carotene for six years did not decrease the risk of AMD compared to placebo (Teikari et al., 1998). A placebo-controlled trial in a cohort of 22,071 healthy US men found that β-carotene supplementation (50 mg every other day) had no effect on the incidence of age-related maculopathy — an early stage of AMD (Christen et al., 2007). Recent systematic reviews of randomized controlled trials have concluded that there is no evidence that β-carotene supplementation prevents or delays the onset of AMD (Evans and Henshaw 2008; Evans 2008).

Osteoporosis

A recent experimental study found that the administration of β-cryptoxanthin to ovariectomized mice limited bone resorption by inhibiting osteoclast differentiation but had no effect on osteoblast-driven bone formation (Ozaki et al., 2015). Yet, evidence of a protective role of β-cryptoxanthin — and other individual carotenoids — against bone loss in humans is scarce. In the prospective Framingham Osteoporosis

Study, data analysis of 603 participants found no association between β-cryptoxanthin intake and changes in bone mineral density (BMD) over a four-year period (Sahni et al., 2009). In contrast, intakes of total and individual carotenoids (β-carotene, lycopene, and lutein/zeaxanthin) were positively associated with proximal femur BMD in men over a four-year period, while lycopene intake was positively linked to lumbar spine BMD in women (Sahni et al., 2009). The 17-year follow-up of participants in the Framingham Osteoporosis Study (946 participants) showed those in the highest tertile of total carotenoid intake (median intake: 23.7 mg/day) had a 51% lower risk of hip fracture compared to those in the lowest tertile (median intake: 7.3 mg/day) (Sahni et al., 2009b). In a much larger prospective study — the Singapore Chinese Health Study — in 63,257 men and women followed for 10 years, the highest versus lowest quartile of total carotenoid intake was associated with a 37% lower risk of hip fracture in men but not in women (Dai et al., 2014). Dietary intakes of α- and β-carotene — but not of β-cryptoxanthin, lycopene, and lutein/zeaxanthin — were inversely associated with hip fracture risk in men (Dai et al., 2014). Nevertheless, in a small, four-year, Japanese prospective cohort study in 457 adults, the risk of osteoporosis was 93% lower in postmenopausal women in the highest versus lowest tertile of serum β-cryptoxanthin concentration (Sugiura et al., 2012). No such association was reported with serum concentrations of other individual carotenoids. Whether carotenoid supplementation may help prevent bone loss and reduce the risk of osteoporosis in older individuals is currently unknown.

Menopause

It has been suggested that it is important to limit intake of phytoestrogen-rich foods and avoid high-estrogen foods during most of someone's life. However, a large variety of studies has proved these dietary estrogens actually help some women during menopause. Menopause is the period of time in which a woman transitions from her last menstrual cycle, ending fertility. While childbearing potential ends, menopause does not have to mark the end of vitality and healthy sexuality. The biggest drawback to menopause is the unpredictable change in sex hormone levels, namely estrogen, progesterone and testosterone. During perimenopause, some physicians suggest beginning to increase phytoestrogen intake to counteract the effects of the hormonal imbalances women begin to experience and balance hormones naturally. Some research indicates a drastic drop it in hot flashes for women in perimenopause eating a phytoestrogen-rich diet. Another benefit phytoestrogens may offer to menopausal or postmenopausal women is a reduction in bone loss, leading to higher bone density and fewer breaks, when administered in dose-specific measures alongside vitamin D. They've also been shown to regulate iron absorption into the bloodstream, offering mild anti-inflammatory effects and protecting against massive iron level fluctuations (Bayele et al., 2015). As of yet, there is not sufficient evidence to suggest a phytoestrogen-rich diet can totally counter and relieve menopause symptoms, but nothing truly can. The symptoms of this hormonal change can be managed but not completely avoided.

Functional foods for an aging population

The ageing population is on the increase across most of the world. The proportion of the global population over 60 increased from 8.6% in 1980, according to the United Nations, to 10.0% in 2000, and is forecast to rise to around 14% by 2020. This trend is most pronounced in developed nations where living standards and availability of medical care are high, with Japan showing the fastest ageing population.

The over 60s is a growing demographic, and a substantial proportion has a large disposable income. This group also has specific demand patterns with regards to food purchases. They may seek out foods which address medical issues linked to old age, and also foods which cater for the decline in taste sensitivity and the need for simple, convenient foods, which are physically easy to handle and prepare.

Although products targeted at the elderly will be a niche area, there is considerable potential for well-marketed products which target the ageing population. Products that address bone health, vitamin deficiencies, diabetes and other complaints common to the elderly should receive a boost due to a gradual swelling of the potential consumer base. Indeed, the ageing of the population should push up sales of functional food in general as incidence of many illnesses increases with age.

Euromonitor's report titled The World Market for Functional Food and Beverages comments on the huge potential of the ageing population as a consumer demographic, "Not only will this group be large, it will, in many countries, also be wealthier, more active and more experimental than previous generations, and thus will represent an excellent target for functional foods."

Effectively tapping into this is a strong prospect, especially for a nutritionally focused food manufacturer. Consumers of this age group are among the most affluent, but to date food manufacturers have largely failed to meet the potential demand this demographic heralds.

The potential for this market is clear. While youth-focused functional foods, such as sports and energy products, currently dominate the functional foods market, their share of the market is expected to decline in favour of cholesterol lowering products and those which aid intestinal health.

These latter products are likely to come to the fore in line with the growing elderly demographic, which forms a major consumer base for these products. Julian Stowell of Danisco highlights a number of key areas where ingredients can be added to foods to allow them to address the medical issues which are becoming increasingly prevalent as society ages. Foods which tackle diabetes, compromised immune systems, digestive issues, malnourishment, osteoporosis, hypertension and high cholesterol, all of which afflict the elderly, offer considerable growth potential.

Functional foods are an exciting potential strategy to help address the challenges associated with the world's aging population. Before potential benefits of functional foods can be realized for older adults, it is important to consider older adults' attitudes toward functional foods, including their food forms, bioactives and health benefits.

Antioxidants in fish have the potential to influence human health positively (Gormley, 2006). For example, Anguilla japonica and Conger myriaster are popular eel species in the Chinese food industry, and eel is also used in traditional Chinese medicine with reported potential as a functional food. Antioxidants protect the body against oxidation by scavenging biologically toxic reactiveoxygen species. Ekanayake et al. (2005) screened potential antioxidant compounds in the skin and flesh extracts of the two eel species above and all extracts of Anguilla japonica showed dosedependent DPPH (2,4-dipicrylhydrazine) free-radical scavenging, 10 and also significant hydroxyl radical-scavenging activities. The diethyl ether extract of this species showed high superoxidescavenging activity. The authors concluded that Anguilla japonica is rich in heat-stable and non-polar antioxidants. In another study, Noguchi (2003) examined the biological activity in halibut skin and found that an alkaline extract showed inhibition of proliferation of some tumour cells in-vitro. It was concluded that fish skin components have significant biological activity and are valuable for functional foods and drugs.

Understanding existing beliefs about food and health can improve our ability to design and promote functionally enhanced foods that make sense to consumers, but if our aim is a healthier future then we should also consider broader historical contexts and implications. Hasler (2000) describes the emergence of functional foods as a radically new approach to diet among consumers that began in the last decade of the twentieth century. But the changes she observes—consumers increasingly using food as a means to pursue optimal health and wellness—are part of a larger set of changes in the culture and politics of health in the United States that have been underway since the 1970s. 3ough they involve increased effort among individuals to control and mitigate health risks through their eating habits, it is not clear that these changes have been, broadly speaking, "good for us." Functional foods as a product category and a health practice emerge from and participate in the establishment of a particular set of understandings around health that began in the 1970s. Around that time, the American public became increasingly concerned about health and safety in the context of growing awareness of environmental / occupational health hazards (carcinogens in food, air pollution, environmental degradation) and lifestyle hazards (smoking, high cholesterol). Political and corporate resistance to health-related regulations combined with a growing professional and public emphasis on individual health promotion led to the emergence of what Robert Crawford (2006) calls "the new health consciousness." He describes this as an "ideological formation that defined problems of health and their solutions principally, although not exclusively, as matters within the boundaries of personal control." For the middle class in particular, the pursuit of health has become a major focus of attention, activity and concern. Understood as an outcome of individual effort and calculation, health also is an increasingly significant marker of self-control and responsibility. Crawford (2006) warns that, as the tasks related to maintaining personal health proliferate, awareness of the social aspects of health recede, leaving “little room for responsibility beyond the quest for personal well-being.” Functional foods are an artifact of this new health consciousness, one of many strategies that twenty-first-century Americans can use to manage an ever-increasing sense of health hazards and to pursue health through individual effort and calculation. Functional foods Biltekoff 105 play a role in increasing the required breadth of health knowledge (by introducing terms such as plant sterols and omega-3s into the public lexicon, for example) and add new tasks to an already extensive list of responsibilities. In doing so, they may participate in the foreclosure of our ability to see beyond individual responsibility and the marketplace of goods as a source of health and to recognize and act proactively in regard to the social factors that shape and impede health. While the aim of functional foods is to promote health, they also add to an ever-increasing sense of health risk; “Am I getting enough antioxidants?” was not an issue that most people worried about before products bearing antioxidant health claims hit the shelves. What Crawford (2006) calls the “anxiety / control spiral” (the more we seek to control risk, the more we know about risk, the more anxious we become about risk) may drive people to buy functional foods, but it may at the same time undermine a sense of security that is also an important aspect of well being. An expansive view of the future of dietary health will include functional foods, but not to the detriment of social, political and legislative remedies that address the structural determinants of health. A better understanding of the ways that consumers negotiate functional foods on the intimate level of consumption in the context of existing values and beliefs about food and health is crucial to offering solutions in the marketplace. But we should not lose sight of the aspects of health that lie beyond consumption and beyond the reach of individual control, those environing social conditions that set the widest possible constraints for dietary health and its pursuit.

THE CONSUMER AND FUNCTIONAL FOODS

There is clearly a difference between the Western perspective on functional foods and the Eastern perspective. In the West, functional foods are viewed as a revolution and represent a fast growing segment of the food industry. Food companies, drug companies, chemical companies and boutique companies arising from grass-roots research and development efforts are all racing to bring functional foods to market.

In the Orient, in contrast, functional foods have been a part of the culture for centuries. In traditional Chinese medicine, foods that have medicinal effects have been documented since at least 1000 BC. From ancient times, the Chinese have understood that foods have both preventive and therapeutic effects and are an integral part of health, a view that is now being increasingly recognized around the world.

The market is constantly flooded by new food products with added value and benefits for consumers. With so many continuous options becoming available the task of selecting what is truly beneficial from what is lone marketing can be a daunting one. Health claims are often made about functional foods and legislation exists in various countries to protect consumers from misleading claims.

Most researchers in the area of functional foods agree that any food designated as "functional" and ascribed specific health benefits should receive scientific scrutiny before specific health claims are permitted. In countries with a rich cultural tradition of belief in the health-enhancing properties of various foods and food components, many foods are associated with specific health claims but may not have scientific documentation to parallel the long-standing experience with these foods.

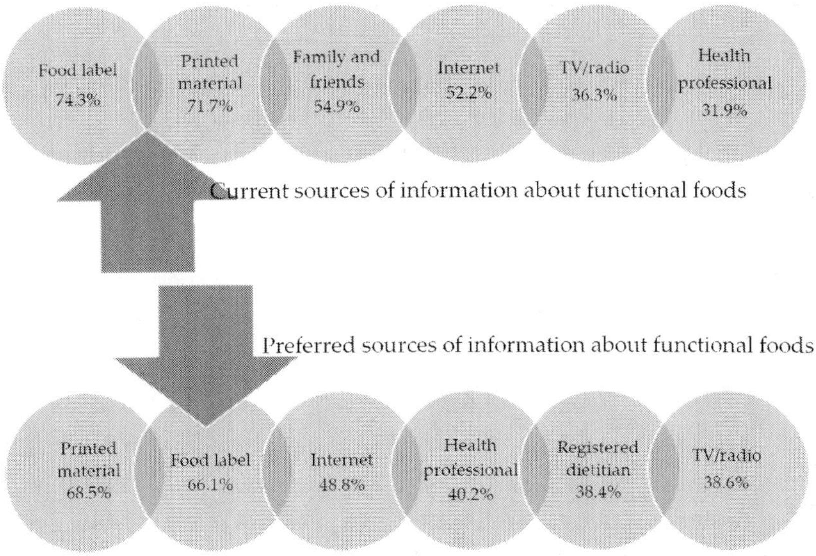

Figure 18: Current and preferred sources of information about functional foods.

Adapted from Vella et al., 2014

In the meantime, the results of the many studies underway will be published not only in scientific journals but likely on the front page of the nation's newspapers. Consumers are eager to act on these findings. Dietetics professionals should continue to recommend a varied, plant-based, balanced diet and follow the studies of specific foods and food components as both treatment and preventative therapy against particular diseases.

Consumer interest in learning more about functional foods remains high. Almost nine in ten Americans (86%) are interested in learning more about foods that have health benefits beyond basic nutrition. Similar to 2011, almost half of all consumers (45%) are "very interested" while another 42 percent are "somewhat interested." (IFIC, 2013). Overall, nearly two-thirds of consumers believe that functional foods will provide health benefits (63%) and have confidence they could make the necessary changes to integrate functional foods into their diet (62%). Breakfast is the preferred meal or snack time when consumers most often choose to include a functional food, with 52 percent of consumers listing breakfast as their first choice among other meal and snack time options. In regards to specific sub- populations who may benefit from functional food consumption, consumers think of infants and toddlers as the age group that would most benefit, followed by children (ages 4 to 12), and people age 60 or older. The success of these products depends on consumer responsiveness. Consumers are becoming health-conscious and most agree that eating healthfully is a better way to manage illness than using medication. Market analysts report that this 'do-it-yourself doctoring' or 'personalised nutrition' is an increasing trend. It is now seen that consumers select foods which they perceive will help them treat or manage their individual health conditions, such as heart health, allergies, arthritis and general wellbeing. Any food product which is promoted as a means of achieving general health and wellbeing or managing or preventing specific health conditions can therefore be considered a functional food.

Market research provides a useful insight into consumers perceived health and lifestyle needs. However, it is equally important to understand public health issues and trends as well as recommendations from health authorities surrounding specific health issues. Food products which are consistent with public health goals are more likely to be recommended by health practitioners and hence are more likely to be sustainable. Targeting key public health concerns provides access to a potentially larger and well defined audience.

For functional foods to be accepted and adopted by both health practitioners and consumers, they will require a strong underpinning in science. Consumers are becoming more discerning regarding food choices. As information becomes more accessible, there will be a need for authoritative scientific evidence which 'proves' that consumption of the food product will provide the promised health benefit. Shifting the focus to key health issues (identified by both consumers and health authorities) and demonstrating the contribution of Functional Foods to positive health outcomes is likely to maintain the credibility and ultimately the sustainability of the functional foods market.

The number of approved health claims in food is extremely low in Europe. Until March 2014, the European Food Safety Authority (EFSA) has only approved 254 health claims out of the 2,242 submitted. This has made Europe a hard market for manufacturers, but also an important one as it provides a gold standard test for health claims.

On the other end of the spectra, there is a country like Japan, whose stable FOSHU system, as well as its less strict demands for clinical evidence, has made the country a popular test market for new ingredients. By early 2014, total approvals in Japan reached over 1,100 – five times higher than the number of approvals in Europe. Japan stands out among developed countries for the range of approved innovative health ingredients.

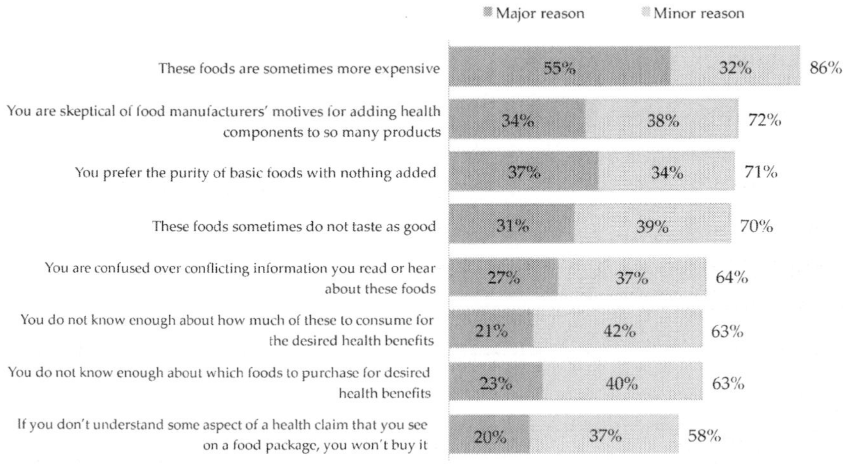

Modified from IFIC 2013

Figure 19: Common barriers to functional food consumption.

A thorough understanding of laws and regulation within and among key countries with regard to functional foods is critical to the direction of food companies that are developing products for these markets.

Functional food cannot be a single well-defined/well characterized entity (Roberfroid, 2002). The concept of Functional Foods has been adopted differently by the countries across the globe. The differences lay in the definition and thereby the scope and the regulatory framework that governs the market. The concept of functional food was first promoted in 1984 by Japanese scientists who studied the relationships between nutrition, sensory satisfaction, fortification and modulation of physiological systems (István et al., 2008). There it referred to the food products fortified with special constituents that possess advantageous physiological effects (Hardy, 2000; Stanton et al., 2005). Functional foods also offer the perception that the food industry is removing less desirable elements in foods (fats, sugars, sodium, etc.) and adding more sought-after ingredients (Mulry, 2002). Another research says that a functional food is, or appears similar to, a conventional food. It is part of a standard diet and is consumed on a regular basis, in normal quantities. It has proven health benefits that reduce the risk of specific chronic diseases or ill states in addition to its basic nutritional functions (Doyon and Labrecque, 2008).

America

The perspectives of Latin America as a potential producer and consumer of functional foods will depend largely on the level of information and income of the population, credibility of the products, research investments and regulatory practices. The characteristics of Latin America are diverse at the regional and sub-regional levels. However, as part of the demographic and epidemiological transition currently underway, common trends can be identified such as increasing urbanization and life expectancy, the occurrence of obesity and malnutrition, increasing incidences of chronic diseases and causes of mortality rates, all of which suggests the importance of diet and functional foods in public health policies. The Latin American population in general has no knowledge of functional foods, but in the more urbanized areas there is an increasingly health-conscious consumer, aware of the importance of food for health, due to the media and local traditions. More investment in research is important to explore the existing plant biodiversity that is a rich source of new foods and bioactive compounds, some of which are already used for health improvement and well-being. Clinical validation of functional foods should consider functional food science concepts and also the diverse cultural and genetic background of the local population. In Latin American scientific and regulatory communities, the functional foods concept has been associated with foods having health benefits beyond those of basic nutrition but is not defined officially in the emerging regulatory codes. Regulation existing in some countries is focused on safety and efficacy; both functional and health claims are allowed (risk reduction) provided they have scientific validation. This allowed introduction on the market of several products with health claims, some of them submitted to post-marketing surveillance.

In the United States, the Food and Drug Administration regulates the claims that manufacturers can make about functional foods' nutrient content and effects on disease, health or body function. The 1990 Nutrition Labeling and Education Act allows, for the first time, health or disease prevention claims on a food label. Within this context the FDA requires that health claims be supported by solid research and that the claims be approved by the FDA. To date only a limited number of claims have been approved, and we're probably several years away from having the necessary research that will support or disprove suspected health benefits of functional foods.

While the FDA of the USA recognizes dietary supplements and medical foods, the USDA does not have a formal definition for "functional food." This gap in U.S. federal legislation creates a challenge for researchers and developers who want to sell or educate the public about functional food. In spite of this, the FDA recently released guidelines for assessing health claims, entitled the "evidence-based review system for the scientific evaluation of health claims" (FDA, 2013). Like Japan and Europe, the U.S. would like to systematically review health claims so that the meanings of health claims are clearer (Ellwood et al. 2009; Lupton, 2009; FDA, 2013). Under the evidence-based review system, food manufacturers may issue "nutrient content" claims, "health" claims, or "structure/function" claims. The former two are similar to the EU's "nutritional" and "health" claims respectively (Ellwood et al., 2009; Lupton, 2009; Verbeke et al., 2009; FDA, 2013). A "nutrient content claim" describes the actual components of a food product without any reference to health. An example of a nutrient content claim would be, "This product contains 1 gram of sugar." A "health claim" explicitly or implicitly ties the food product to health or disease. However, a health claim may only signify risk reduction of a disease and not "diagnosis, cure, mitigation, or treatment of disease." An example of a health claim would be, "This product promotes cardiovascular health." Finally, a "structure/function" claim simply relates a nutrient to the structure or function of the human body based on accepted scientific research. An example of a structure-function claim would be, "This product contains Vitamin A. Vitamin A is important for maintaining healthy eyes." Finally, the FDA allows "qualified health claims" on food products whose effects have been suggested but not verified. Nevertheless, food manufacturers must include a disclaimer on the indicate product indicating that the scientific evidence behind the health claim is weak (Lupton, 2009; Verhagen et al., 2010; FDA, 2013). In order to verify food claims, the FDA pays special attention to research involving the administration of food compounds (at specific dosages) to animals and humans (Ellwood et al., 2009 ; FDA, 2013). Claims must then be approved under "Significant Scientific Agreement" after which FDA officials rank claims based on the strength of the scientific evidence behind them.

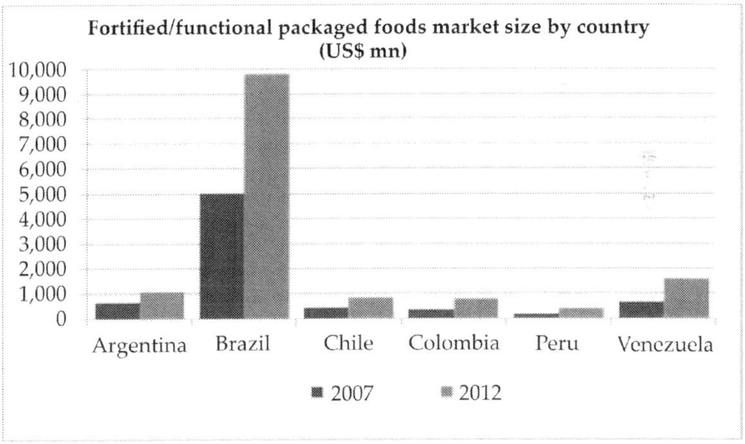

Adapted from Euromonitor International

Figure 20: Functional and fortified food market size by country in South America.

In Latin America and the Caribbean, regulation concerning functional foods, functional supplements and functional and health claims differs greatly from country to country. There is no official or legal definition of functional foods or a specific regulation for functional foods/ingredients. In general, basic nutrient content/function claims are allowed and subject to some norms, but only a few countries allow or have norms for health claims. Even so, health authorities have allowed product claims by on a case-by-case basis in several countries. Brazil is the only country to have a well-defined regulation for functional and health claims for either nutrient or non-nutrient components and also for the demonstration of safety and efficacy (Agencia Nacional de Vigilancia Sanitaria, 1999). In almost all countries functional or health claims associated with non-nutrient compounds or non-essential nutrients are neither prohibited nor regulated. As a consequence, there is no legislation on efficacy and safety and also a lack of criteria for scientific substantiation of claims, except for Brazil. Even so, all over Latin America, in popular markets and even in supermarkets, hundreds of foods and drugs are sold together with curative allegation. Considering the regional economic level, cultural identity and public health situation, the development of common functional foods concepts and norms seems to be very important. Brazilian legislation can be used as a starting point in this direction since it has been based on internationally acceptable principles and is considered to be up-to-date. In Brazil functional foods have not been defined as such but the norms were based on the idea of a food that is food and not a drug, that is part of a normal diet and that can produce benefits beyond basic nutrition. Legislation essentially rules on demonstration of the safety and efficacy of novel foods (foods not used before in Brazil) and foods/ingredients that have a claim in the label. All of these products should be registered and approved by health authorities.

Europe

When functional food science migrated to Europe, researchers defined "functional food" using the following statement:

"Food products can only be considered functional if together with the basic nutritionalimpact it has beneficial effects on one or more functions of the human organism thus eitherimproving the general and physical conditions or/and decreasing the risk of the evolution ofdiseases" (Siro et al., 2008; Diplock et al., 1999).

As od 2015, the European government categorizes food into: "conventional foods, modified foods, foods for special dietary use and medical foods" (Cara, 2014). Unlike Japan, however, the EU government does not have a formal legislative definition for "functional foods." According to the European Commission Concerted Action on Functional Food Science in Europe (FUFOSE) and PASSCLAIM, food developers are permitted to make two types of claims: nutritional and/or health claims. "Nutrition claims" refer to a food's basic nutrient content and ability to provide energy. "Health claims" refer to a food's ability to prevent, manage, or treat illness. These titles, particularly "health," must be backed by significant scientific evidence, similar to Japan's FOSHU application process. The steps are (Lee and Foo, 2014; Zawistowski 2008; Clydesdale, 19997): requirements for health claims are stringent.

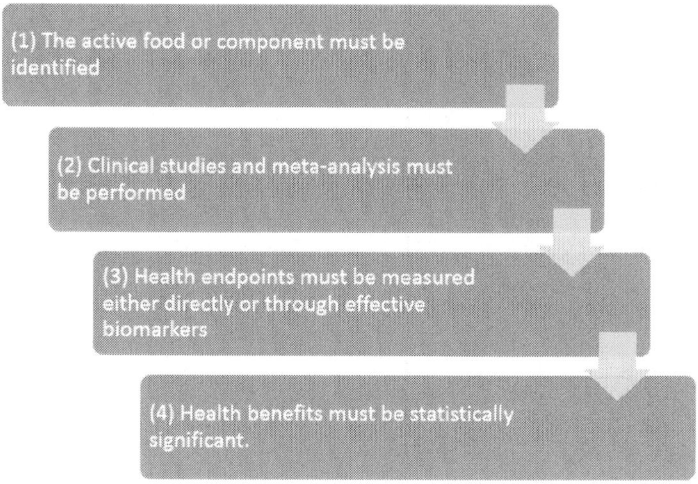

Figure 21: Steps towards establishing health claims.

As a result, they bar many food developers from printing health claims on their proposed "functional foods." Moreover, functional food developers find extensive testing costly. In other words, the lack of a clear, cohesive definition may be leading to over-regulation of health claims in the EU, impeding functional food development there.

The fact that the European markets for functional foods generally are less developed, compared to the US and the Japanese markets, has often been attributed to a restrictive and inconsistent health claim legislation in and between the European countries. With the European Parliament's second reading of the main principles of the harmonized regulation COM/2003/0424, this situation is changing.

Within the European Union (EU), a regulation on nutrition and health claims came into force in 2007. Under the regulation, health claims are subject to pre-approval, involving scientific assessment by the European Food Safety Authority (EFSA). However, although EFSA assesses the scientific dossiers supporting potential health claims, the final decision to accept or reject a claim lies with the European Commission.

The European Commission, through its 5th Framework Programme, began focusing on a substantial effort in the science of the intestinal microbiota, its interaction with its host and methods to manipulate its composition and activity for the improvement of human health and well being. Eight multicentre and multidisciplinary research projects now cover a range of topics required for the development of efficacious probiotic foods, from understanding probiotic mechanisms at a molecular level; developing technologies to ensure delivery of stable products; and demonstrating safety and efficacy of specific probiotics in defined treatment targets (Saarela et al., 2002).

Because of increasing interest in the concept of "Functional Foods" and "Health Claims", the European Union set up a European Commission Concerted Action on FunctionalFood Science in Europe (FUFOSE). The programme was coordinated by the International Life Sciences Institute (ILSI) Europe and the aim was to develop and establish a science-based approach to the evidence needed to support the development of food products that can have a beneficial effect on an identified physiological function in the body, that can improve an individual's state of health and well-being and/or reduce the risk of disease. The FUFOSE project looked at six areas of science and health: growth, development and differentiation, substrate metabolism, defence against reactive oxidative species, functional foods and the cardiovascular system, gastrointestinal physiology and function, and the effects of foods or behaviour and psychological performance. The final document was published in the British Journal of Nutrition.

The report takes the position that functional foods should be in the form of normal foods and they must demonstrate their effects in amounts that can normally be expected to be consumed in the diet. A functional food can be a natural whole food, a food to which a component has been added, or a food from which a component has been removed by technological or biotechnological means. It can also be a food in which the nature of one or more components has been modified, or a food in which the bioavailability of one or more components has been modified, or any combination of these possibilities. A functional food may be targeted at the whole population or for particular groups, which may be defined, for example, by age or by genetic constitution.

The EU Concerted Action supports the development of two types of health claims relevant to functional foods, which must always be valid in the context of the whole diet and must relate to the amounts of foods normally consumed.

These are:

- TYPE A: "Enhanced function"claims that refer to specific physiological, psychological functionsand biological activities beyond their established role in growth, development and other normal functions of the body.

This type of claim makes no reference to a disease or a pathological state, e.g. certain non-digestible oligosaccharides improve the growth of a specific bacterial flora in the gut; caffeine can improve cognitive performance.

- TYPE B "Reduction of disease-risk"claims that relate to the consumption of a food or food component that might help reduce the risk of a specific disease or condition because of specific nutrients or non-nutrients contained within it (e.g. folate can reduce a woman's risk of having a child with neural tube defects, and sufficient calcium intake may help to reduce the risk of osteoporosis in later life).

Asia

There is growing interest by the food and supplement industries to expand into the Asian marketplace because of the large market size, the Westernization of lifestyles, and the long-standing history of using foods and natural products for health promotion. With these new opportunities comes the challenge of navigating the various regulatory landscapes. Japan is credited with pioneering the term "functional foods" in 1984, later replaced by the term "health foods." One category of health foods that is of great interest to ingredient companies is Food for Specified Health Uses (FOSHU), which refers to foods containing ingredient(s) that are consumed for the maintenance or promotion of health, for the control of health conditions (e.g., blood pressure or blood cholesterol), or for disease risk reduction. Evaluations of FOSHU are based on scientifi c substantiation of effi cacy, safety, and quality. The requirements for demonstrating effi cacy and safety in a FOSHU application are stringent and require studies to be conducted on the food containing the ingredient.

The Chinese population represents over 1 billion potential customers. Companies seeking to introduce ingredients that do not have a significant history of consumption in China (i.e., novel or new resource foods) are required to obtain regulatory approval by the Ministry of Health (MOH). The approval process involves an evaluation of the quality and safety of the ingredient by an Assessment Committee. The Committee also determines the need for additional testing in support of safety; this testing must be conducted by a testing agency accredited by the MOH. For those intending to include health claims in the labeling and advertising of their product in China, registration of the product as a health food is required. Information related to the quality, effi cacy, and safety of the product is evaluated by the State Food and Drug Administration (SFDA). Health foods must undergo testing by an SFDA-designated agency before a health food application can be submitted. In South Korea, "health/functional foods" (HFFs) contain nutrients or other substances with a nutritional or physiological effect. HFFs are divided into two subcategories: generic and product-specifi c. Products that are included in the established list of generic HFFs do not require pre-market approval, whereas all new ingredients that are not considered generic HFFs (i.e., product-specifi c HFFs) must undergo premarket evaluation by the Korean Food and Drug Administration. Evidence of the standardization, safety, and effi cacy of the product must be provided. As the regulatory requirements for HFFs are stringent, only highly substantiated ingredients receive approval. For conventional foods containing functional ingredients, authorization is granted on an individual basis. Applicants must submit a dossier containing information demonstrating the safety and functionality of the fi nal product. Companies seeking to market new ingredients in South Korea must meet the requirements of the Novel Food regulations, which were implemented in 2010.

The Indian food industry is no different from the western counterparts in its response to the growing need for healthy food. With the fast changing socio-economic indicators, the Indian consumer is now looking for the food products that provide value beyond nutrition. This value may be the specific health benefits, the disease risk reduction ability, the general well being or the cosmetic benefit associated with the foods. The functional foods are particularly targeted with such value propositions and are now increasingly been accepted by the consumers across the food categories. But still the functional foods in India have a long way to before establishing as a mature market offering. The deteriorating state of health in urban India raises questions about the adequacy of the conventional food products to address to gasping issue. The problem, although multi-dimensional in nature, may partially be addressed by functional foods introduced in the Indian market.

The Food Safety and Standards Authority of India (FSSAI) has issued its long-awaited rules for functional foods to be enforced from January 1, 2018. It is hoped that the rules will put India on a similar regulatory level to leading nations while lso establishing a strong scoence-based environment for the industry in the country.

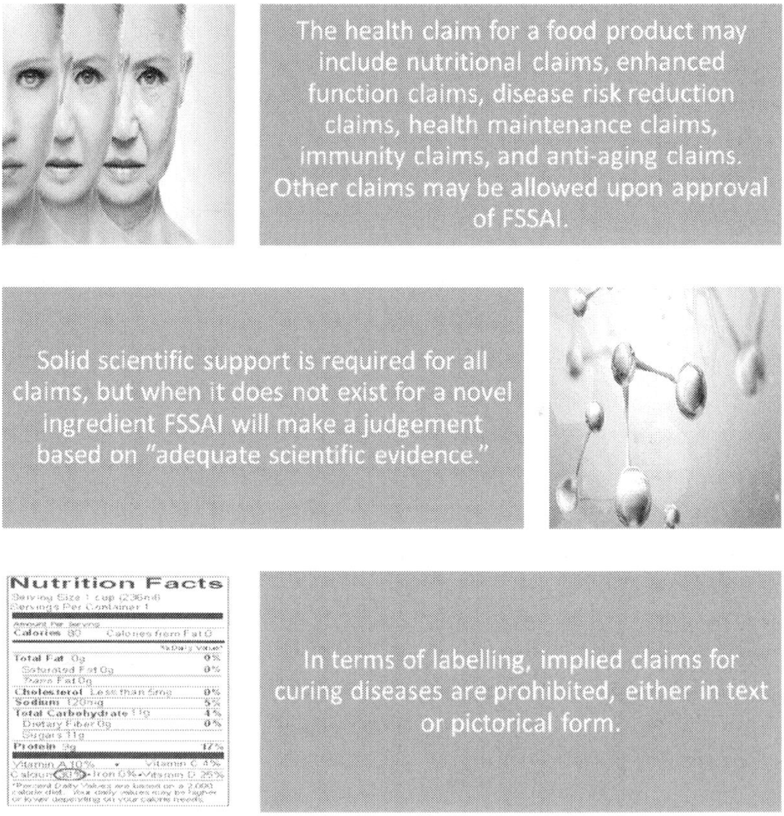

Figure 22: Summary of the rules regarding functional foods.

The Asian market presents the greatest opportunity for companies to expand their marketing and sales potential beyond North American and European borders. Given the complexities and differences associated with obtaining regulatory approval for new ingredients and products in

Asian jurisdictions, it is important to gather a thorough understanding of the specifi c requirements in each country. Developing relationships with local distributors and working with consultants with expertise in Asian regulations also are key to successfully entering the Japanese, Chinese, and South Korean markets.

Safety Issues

In addition to the normal expectation that any food item or ingredient be safe for consumption, functional foods have the additional safety requirements related to the novelty of the raw materials used, of the processes used to produce the food, or of the potential use in the diet. For example, it is very likely that many food components already present in foods will be found to have positive health benefits and will be used in a novel way in the diet, which could increase the amount of these foods consumed each day or might increase the use of these foods by at-risk groups that currently do not use much of these foods or their components. The safety evaluation must then take into account the effect of increased consumption on various age groups and the infirm and ask whether increased use might lead to increased exposure to potentially harmful components, such as natural toxins or pesticides.

When a functional food contains novel ingredients or is produced by a novel process, it becomes particularly important that safety as well as acceptability be considered. With traditional foods, safety assurance comes from a long history of safe use by large populations over a number of generations. Such history is lacking for novel ingredients and novel processes, and a safety assessment is required before such foods are released into the food supply.

A detailed assessment of studies on toxicity, reproductive toxicity, genotoxicity and estrogenicity of phytosterol and phytostanol esters was carried out by the 69th meeting of the Joint FAO/WHO Expert Committee on Food Additives (JECFA) (World health Organisation, 2009). JECFA. JECFA found no evidence for genotoxicity of phytosterols or phytostanols and their esters and no indication of potential for carcinogenicity from the available toxicity studies.

Although randomized, double-blind, placebo-controlled clinical studies are the gold standard for determining ef- ficacy and safety of functional foods, a person can potentially compare results of multiple studies involving functional foods when researchers along with journal editors ensure that a standard checklist of information is reported within each study. Such information should include a minimum of the following items: subject (eg, type, number, and age); use of a control; dosage form (ie, food composition, active compound within the formulated food, source of the active compound); standardized source (eg, compositional verification via chromatographic analysis or some other appropriate testing method, including genus, species, and variety/accession for plants/herbs, or supplier identification); frequency of intake; duration (weeks vs months); and measurable outcomes dependent on quantitative end points. Measurable outcomes must also relate to the hypotheses proposed in the study design when evaluating the efficacy of functional foods. In addition, measurable outcomes or end points should make use of standardized measures of testing or biomarkers. By providing this type of detailed information within each study, comparison and duplication of research results, along with statistical modeling and meta-analyses, can be conducted to fully substantiate the science behind functional foods and bioactive food compounds. Such consistency in reporting would assist in providing the preferred reporting items for systematic reviews and meta-analyses (Moher et al., 2009). This is an evidence-based minimum set of items for reporting in systematic reviews and meta-analyses.

The microbiological or residue aspects of functional foods are taken as read and are not addressed here. However, there are safety considerations associated with the ingestion of bioactives from various functional foods. For example, ω-3 PUFAs occur naturally in fatty fish and food technologists are now adding PUFAs and other bioactives to many foods so there is a danger that consumers could get an overload; this may or may not be harmful. Several models have been developed to calculate safe maximum amounts of vitamins and minerals for the purpose of food fortification. However, none of the models include the adding of bioactive ingredients to foods. At present, the number and variety of bioactive substances used as functional food ingredients is growing rapidly (Niemann, 2010); they are a diverse group of chemicals including complex mixtures and microbes. All these substances are normal constituents of our daily diet in low and/or trace amounts but we run the risk of obtaining oversupply because they have been added to a range of foods. 19 A simple strategy to derive reasonable amounts of functional food ingredients could be the follows: (i) the substance has to be characterised and all traditional foods containing it have to be determined; (ii) the population with the highest habitual intake of these foods has to be selected for calculating the intake of the substance; (iii) the highest identified habitual intake could be divided by a factor of 10 to consider the multiple exposure by all the other food commodities containing the same ingredient. The resulting amount could be added to one serving size of a given food product. However, this approach cannot be used for functional foods claiming dose-related effects that require amounts beyond all nutritional experience (Niemann, 2010).

Regulatory Issues

Functional foods exist at the interface between food and drugs. There is no provision in our existing food regulations for foods intended to be consumed to prevent disease. As things stand, dietary supplements will be marketed under the Dietary Supplement Health and Education Act, and foods will be marketed under the Nutrition Labeling and Education Act.

The concept of "substantial equivalence" was developed as a practical approach for evaluating the safety of foods that contained or were produced from genetically modified organisms. This approach asks whether the novel food is substantially equivalent or sufficiently similar to its traditional counterpart. This concept was not intended to be applied to other types of novel foods but appears to be increasingly used by various regulatory agencies. The idea is that, by focusing on what is different between a novel food and a traditional food, the safety assessment can zero in on the differences and thereby minimize the expense and time required for the assessment. An International Life Sciences Institute-Europe technical committee established three different classes of equivalence and set up guidelines for safety assessments of novel foods as described on the figure below.

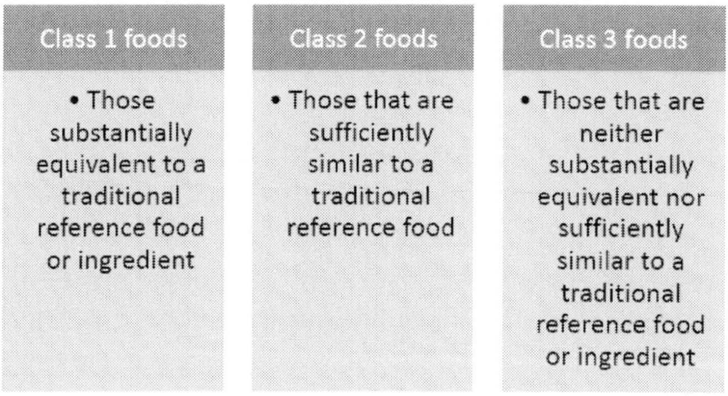

Figure 23: Classification of foods and food equivalents in accordance to the International Life Sciences Institute-Europe.

In order to establish the appropriate equivalency class, extensive background information about the food is required, which ranges from the raw material source and its origins to methods of production and expected use of the novel food. Class I foods would not be subject to safety assessment, Class II foods would be but the testing would focus on the differences between it and the traditional food or ingredient, and Class III foods would be subjected to rigorous safety assessment on a case-by-case basis.

Assuming the food or ingredient passes the rigorous safety considerations, each health claim will need to be subjected to scientific evaluation as well, which could take years to carry out. In the meantime, consumers will become increasingly interested in using functional foods, and we need to stay on top of developments in this area.

While there is legislation in several countries for plant sterols in oils and spreads (eg the USA, Canada, Switzerland and EU) other dietary sources are not so common. However, the EU Scientific Committee on Food (SCF) considered that phytosterols from multiple dietary sources should avoid plant sterol intakes exceeding 3 g/day (SCF, 2002a). The Scientific Panel on Dietetic Products, Nutrition and Allergies of the European Food Safety Authority (EFSA) recommended that sterol-containing foodstuffs should not be consumed in amounts resulting in total phytosterol intakes exceeding 3 g/day.

The largest number of functional food and drinks consumers is in the United States – especially as a market for lifestyle health products.

Emerging hubs such as China, India and Brazil are less important for now, but stand to exert significant influence over the functional food and drinks market in the future.

The future of the regulatory landscape is hard to predict. Certainly, large emerging markets such as India, China, and Brazil will adopt stricter regulation, although it is unclear how closely this will be enforced in the short term. The second key future event is the 'rump' of approvals in Europe – particularly the possibility of approval for probiotic health claims.

Labelling schemes

There is a call from the scientific community for proper regulation and food labelling to ensure consumers clearly understand the safe and appropriate upper and lower limits. Individuals are consuming products that claim to be functional foods without knowledge of what they are taking at times. In various countries, including EU member states, Australia and New Zealand, added nutrients for the development of functional foods must be declared in the statement of ingredients on foods to enable consumers to identify these added ingredients. While the requirements vary from region to region and country to coutry, it is generally considered good practice by manufacturers and has often demanded by informed consumers. For example, approval to use phytosterol esters derived from vegetable oils first came into effect in Australia and New Zealand in mid-2001 but limited to edible oil spreads. Plant sterols were subsequently permitted to be added to certain edible oil spreads, and certain breakfast cereals, milk and yoghurt (after two reviews requested by Ministers) in 2006.

Health Claims

A health claim is a claim made on food that states, suggests or implies that a relationship exists between a health benefit and a food category, a food or one of its constituents. It is important to assess each individual food on its own merit. In particular, it is crucial that robust science exists to underpin the claims being made. Claims about health benefits of functional foods must be based on sound scientific criteria (Clydesdale, 1997). Typically, this will include evidence that the substance is absorbed or reaches its site of action; that consumption of the food beneficially influences a physiological function in humans (e.g. blood pressure) or a biomarker recognised to impact on health (e.g. blood cholesterol) and, ideally, that this effect has a direct impact on health status.

When discussing functional foods and the consumer, safety is a critical issue. The optimal levels of the majority of the biologically active components currently under investigation have yet to be determined. The benefits and risks to individuals and populations as a whole must be weighed carefully when considering the widespread use of physiologically-active functional foods. For example, what are the risks of recommending the increased intake of compounds (e.g., isoflavones) that may modulate estrogen metabolism? Knowledge of toxicity of functional food components is crucial to decrease the risk: benefit ratio.

The health claim of a functional food may be implicit ("rich in vitamin C"), or vague ("strengthens the body's defence system"), but invariably the product is presented with the suggestion of a benefit. Sales of such products are huge and growing. What ingredients do such foods contain—and who safeguards the truth of claims?

Many functional foods contain added vitamins, minerals, and other essential nutrients. Some of these added nutrients indeed promote health: folic acid reduces the risk of neural tube defects, table salt with potassium reduces blood pressure, and polyunsaturated fatty acids reduce the risk of heart disease. But other claims are more dubious—for example, that zinc lozenges protect against colds or that drinks rich in vitamin C protect against cardiovascular disease.

Functional foods may also contain non-nutritive ingredients. Examples of effective non-nutritive ingredients are sugar alcohols in chewing gum, which reduce risk of dental caries; plant stanols and sterols, which lower low density lipoprotein cholesterol (although effects on heart disease remain to be shown); and probiotic bacteria, which may diminish rotavirus diarrhoea in infants. But other effects of probiotics are insufficiently substantiated, as are effects of phytoestrogens against breast cancer (Kurzer, 2003) of oligosaccharides for "gut health," of flavonoids against heart disease, and of conjugated linoleic acid for weight loss. Herbs such as Kava, St John's wort, and echinacea can also be considered non-nutritive ingredients. They are sold as supplements and added to foods, but their efficacy is controversial and concern remains over potential harm (Tesch, 2003).

Functional foods are marketed directly to consumers, who are unable to assess the implied health claims. Consumers thus must rely on their governments to make sure that they are not misled. Unfortunately, current government regulations leave room for misleading claims. Deception is promoted by the fact that legislation of health claims for foods is layered: there are soft claims, which require soft evidence, and hard claims, which require harder evidence. For example, a draft regulation of the Codex Commission of the United Nations, which sets international food standards, recognises claims about nutrient content ("rich in calcium") and disease reduction ("prevents osteoporosis"), as well as various intermediate categories. Manufacturers have therefore made the formulation of soft claims into a fine art, creating claims that imply health effects without actually naming a disease.

Regulations differ between countries. Japan was the first country to recognise functional foods as a separate category when in 1991 it introduced the FOSHU (Foods for Specific Health Use) system to evaluate health claims. This system has valuable aspects: it regulates both safety and health, and it demands that the food be analysed for the amount of effective component (Shimizu, 2002). But it is voluntary, and even though the evidence required has been reduced in recent years and is minimal by pharmaceutical standards, most manufacturers opt for softer categories of claims, which require little evidence. An example is the unproved but lawful statement that extra vitamins help to maintain healthy skin and mucosa. Watering down of regulations has also occurred in the United States, which once had a solid system for disease reduction claims for foods, which were allowed only if there was "significant scientific agreement" that the claim was valid (USDA, 2003). However, the Food and Drug Administration's oversight over health claims has eroded, and the United States now allows "qualified health claims" for which there is hardly any evidence, as long as a disclaimer is included. In the European Union the safety of novel foods is thoroughly regulated but health claims are not—EU legislation for nutrition claims is complex, fragmented, and poorly enforced. Paradoxically, current EU regulations prohibit claims that a food ingredient prevents a disease even when the claim is true—for example, that folic acid prevents neural tube defects. Finally, Canada, Australia, and New Zealand have introduced new systems to regulate health claims, but experience with these is still limited.

The lack of proper regulatory oversight has led to some functional foods that are no more than quackery, while at the same time other functional foods do promote health and prevent disease. The potential for effective functional foods is certainly there. Foods and food components could prevent or ameliorate many diseases (Hu, 2003 and HCN, 2003), but not enough research is being done to identify effective ingredients and substantiate their efficacy and safety. Whether such research will be done depends to a large extent on proper regulation. Major food companies are eager to expand into health promoting foods, but there is no incentive to underpin such health effects with solid research when products can be successfully marketed on the basis of vague allusions alone. But there is hope. After more than 20 years of deliberations the European Commission recently agreed on new regulations that would prohibit vague claims and that would allow hard claims of disease reduction for foods if the evidence is solid (CEC, 2003). The commission even wants to grant companies seven years of exclusivity for truly novel claims backed up by solid data. If the European Parliament accepts these proposals it would be a step in the right direction.

Table 22: Nutrition Labeling Education Act approved health claims

Calcium & osteoporosis
Sodium & hypertension
Dietary fat & cancer
Dietary saturated fat and cholesterol & risk of coronary heart disease
Fiber-containing grain products and fruits and vegetables & cancer
Fruits, vegetables, and grain products that contain fiber, particularly soluble fiber, & risk of
coronary heart disease
Fruits and vegetables & cancer
Folate & neutral tube defects
Dietary sugar alcohol & dental caries
Dietary soluble fiber & coronary heart disease
Soy protein & coronary heart disease
Plant sterols and plant stanol esters & coronary heart disease

In communicating with consumers about the food/ health relationship, it is important to remember that functional foods are part of an effective strategy for health, not a substitute for good health habits. This is particularly important in light of the numerous health claims and product statements that regularly confront the consumer. Since the enactment of the Nutrition Labeling Education Act (NLEA) in 1990, only specific health/ disease claims have been authorized for use on conventional foods and dietary supplements based on the U.S. Food and Drug Administration's (FDA) determination that significant scientific agreement must exist for such claims (Deis, 2003; Lucchina, 2003). An example of a claim relating soy protein intake with risk of coronary heart disease could read, "25 grams of soy protein a day as part of a diet low in saturated fat and cholesterol may reduce the risk of heart disease. A serving of ___ supplies ___ grams of soy protein." Also, based on the passage of the Dietary Supplement Health and Education Act (DSHEA) of 1994 and a related act in 1997, foods and dietary supplements may communicate "structure/ function" claims and "authoritative statement" claims linking a product's ingredients with health outcomes. For example, stating that consumption of a food product "helps to maintain normal cholesterol levels" would qualify as a structure/function claim, for which prior FDA approval is not required. On the other hand, communicating that this product "lowers cholesterol levels and reduces heart disease risk" is a health claim for which FDA approval is mandated. Of considerable interest is the announcement in late 2002 by the FDA of its Initiative on Consumer Health Information for Better Nutrition (FDA 2003), a plan intended to improve consumer access to scientifically sound and understandable information, thus assisting consumers in making appropriate diet choices, and stimulating competition within the food industry toward developing and marketing healthier food products. A component of the initiative involves expanding the current authorization system to include "qualified" health claims based on a consensus of qualified experts after examination of the quality and consistency of scientific evidence (Meister 2002). The highest level of health claim ("A") would continue to be reserved for those statements meeting the current FDA standard of significant scientific agreement (an unqualified health claim). An example of a "qualified" health claim for nuts could read "Scientific evidence suggests but does not prove that eating 1.5 ounces per day of most nuts, such as walnuts, as part of a diet low in saturated fat and cholesterol may reduce the risk of heart disease."

Despite the emerging interest and global consumption of functional foods, the recent opinion of European Food Safety Authority (EFSA) on article 13.1 for health claims has caused considerable concern and consternation among nutritionists and food manufacturers alike (Hughes, 2009. Of the 416 claims evaluated, fewer than 2% were approved, a rejection rate that calls into question the process itself. Claims related to antioxidants, bowel function, neurological function and glycaemic index were rejected. The rejection of carbohydrates and their glycaemic indices appears to be at variance with other international bodies. The World Health Organization, for example, not only recognizes the concept of glycaemic index, it advocates its wide usage (FAO/WHO, 1998. Yet EFSA states that 'carbohydrates that induce a low/reduced glycaemic response and carbohydrates with a low glycaemic index (<55), which are the subject of the health claims are not sufficiently characterised'. This is hard to reconcile as the carbohydrate content of several foods has not only been carefully characterized but also recognized as being metabolized and absorbed differently. Even more curious is the approval of β-glucan for lowering cholesterol but not for lowering blood glucose. These anomalies highlight the challenges ahead. EFSA urgently needs to balance scientific judgement and consumer protection with the promotion and nurturing of food innovation within Europe. There is an urgent need for EFSA and the food industry to recalibrate each other's expectations. If not, EFSA will be perceived as a harbinger of doom for the industry and scientists alike, stifling creativity and ultimately the consumer will be the loser.

Ethical and sustainability Considerations

Many consumers are aware of ethical and sustainability issues relating to the foods they purchase. This applies equally to whether they are buying unprocessed foods (inherently functional) or food products which may have added functionality as well. With the development of all new functional foods come new ethical considerations regarding the understanding of food consumption.

Among existing understandings about food and health that play roles in how functional foods are negotiated and invested with meaning by consumers, the "health-pleasure paradox" is particularly significant (Biltekoff, 2010). This is the long-standing idea that "good" foods taste bad and good-tasting foods are "bad for us." Mark Twain (1897) expressed this paradox beautifully in Following the Equator: "The only way to keep your health is to eat what you don't want, drink what you don't like, and do what you'd rather not." The sentiment endures because it expresses beliefs about the meaning of health that are prevalent in Western, individualist cultures. We see health through the lens of our social and cultural values, imbuing the pursuit and accomplishment of health with many of the same qualities that have historically been considered essential to mature subjectivity and good citizenship: autonomy, willpower and self-restraint (Crawford, 1984). To the extent that we see health as an accomplishment that is achieved through the exercise of the thinking self over bodily desires, we also understand health as anathema to carnal pleasure and associate healthy practices with self-denial. One of the allures of functional enhancement is that it has the capacity to remake "bad food" as "good for us" and, perhaps, engineer a solution to the health-pleasure paradox. By combining the hedonics of a sinful treat with the nutrient profile of a "good food," products like Vitamuffins™ and Super Donuts® seem to make it finally possible for Americans to indulge their way to health. While these products may indeed offer both pleasure and health, they are negotiated within the context of the health-pleasure paradox and understood by consumers in ways that clearly distinguish the pleasures that they offer from muffins and donuts that have not been enhanced with fiber, vitamins or essential fatty acids and packaged in calorie-controlled portions. 3e pleasure of the functional treat is what Coveney and Bunton (2003) call "disciplined pleasure"…"reasoned, reasonable and safe," rather than risky or unplanned, and consistent with, rather than opposed to, the self-restraining dictates of health. 3is is not the reckless pleasure of a "sinfully" delicious dessert, but a studied attempt to achieve a balance between enjoyment and responsibility that is simply inconsistent with the experience associated with unrestrained—and therefore often guilt-inducing—gustatory pleasures. A fan of Deep Chocolate Vitalicious® VitaTops is quoted in the "Tastymonials™" section of the company Website[2] raving about the disciplined pleasures that the product affords: "When I eat the chocolate ones I feel like I am cheating on my diet when in fact I am eating something really healthy…I thank you for helping me start my day off with something that looks and tastes sinful but is great for me." Not surprisingly, in her study of Finnish consumers, Niva (2007) found that "functional foods had no place in indulgence. 3ey were excluded since, by definition, they encompass a planned controlled health orientation…" Because the experience of disciplined pleasure is defined against indulgence, it also brings with it the pleasure of moral superiority over those whose unregulated pleasures are taken as a sign of irresponsibility, indifference and immorality (Niva, 2007). Product innovation that aims to use functional enhancement to align eating habits with healthy outcomes should attempt to better understand, motivate and capitalize on the existing drive for disciplined pleasures while bearing in mind the ways in which it is distinct from the pleasure of indulgence.

Functional foods are also negotiated and contextualized in relation to existing consumer understandings about what constitutes a healthy diet. But the appeal of functional foods—with its emphasis on single nutrients and the value of specific foods—conflicts with fundamental tenets of dietary health that have been promoted in the United States for over a century. Since the emergence of the science of nutrition in the late nineteenth century, dietary advice has focused on communicating the basic fact that foods are composed of nutrients and that certain foods are, therefore, nutritionally similar despite their apparent differences. Early twentieth-century dietary reformers sought to teach the urban poor that the protein in a cheap cut of tough meat is the same as that in the most expensive steak, and that the real value of food comes not from its cost but from its nutritional content. 3ese ideas were popularized in a campaign during World War I that promoted the idea that beloved but scarce foods could be replaced by abundant yet unfamiliar ones at no nutritional cost. During World War II, the food-group approach to dietary health was introduced and Americans learned that eating right meant choosing at least one food from each of seven groups every day (Levenstein, 1988, 1993). 3e idea that certain kinds of foods share nutritional qualities with others and that a good diet is varied and balanced remained central to dietary advice throughout the postwar period. As the nutrition and health communities shifted their focus from preventing vitamin-deficiency diseases to managing the chronic diseases of middle age, dietary advice became increasingly focused on warning people to avoid or reduce consumption of potentially harmful foods (sodium, cholesterol, fat, etc.). 3e USDA Food Pyramid's hierarchical structure conveyed the notion that some foods are better eaten in abundance and others in moderation, but remained focused on the dietary totality. In the context of this new dietary paradigm, which Warren Belasco has termed "Negative Nutrition," the food industry—concerned about the potential for negative impact on sales—played a major role in shaping an ongoing "no good foods or bad foods" approach to dietary advice that emphasized the importance of the whole diet rather than the role of specific foods within it (Belasco, 1989; Nestle, 2002) Some observers have claimed that the emergence of functional foods represents a shift from a "food negatives" paradigm to a more positive approach to dietary health that entails seeking out functional benefits in food (Hasler, 2000; Leighton, 2002). While consumers are clearly interested in using foods to enhance health in new ways, the emphasis on the nutritional distinctiveness of particular foods and the particular value of specific micronutrients conflicts with traditional nutritional ideals. Although individual functional-food products may appeal to consumers seeking particular benefits, the general concept of functional foods may confuse consumers and be difficult to integrate into existing understandings of dietary health. In her analysis of lay perspectives on functional foods, Lotte Holm (2003) pointed to two related dangers. She noted that the "detail oriented and fragmented" messages of functional foods are similar to food-safety messages that have historically generated "confusion, uncertainty and ambivalence." She also noted that by introducing elements of one group into foods from another (vitamins in candy, for example), functional foods disturb the logic of food groupings and may, therefore, be "counter-productive to the nutritional health of populations."

With the public backlash around the world, against genetically engineered (GE) food, the biotech companies have attempted to repackage the way they promote such foods, by creating a "second generation" of GE foods. By promoting functional foods, and its potential humanitarian benefits and a solution to world hunger, it helps to repair and improve the tarnished image of many of these biotech companies.

Additional criticisms about the real value and practicality of such foods are discussed below, and furthers the questioning of the motives of the biotech industry -- are they just trying to improve their image, or are they really trying to help? If they are really trying to help, why are they receiving such criticism from the very groups of people that would likely receive the products or be working in poverty alleviation, hunger problems and so forth? The latest such foods are those that attempt to tackle malnutrition.

Sustainability

In times of climate change, growing world populations, and with clear evidence of the depletion or damage to critical resources, it is imperative to take action to sustain and develop the capacity of agricultural and manufacturing systems to continue to provide food, the most basic of human needs. Functional foods, generally defined as foods that provide benefits beyond basic nutrition, represent an important growth category for the commercial sector in many countries around the world. Because of widespread concerns about risks to health, the public is vulnerable to messages about potential health protective advantages of foods formulated in response to emerging nutrition science. This paper argues that it is important to finds ways to evaluate the effects of innovation in this sector. Now more broadly defined, the concept of sustainability provides guidelines for decision-making and a framework for assessing the impact of the development of functional foods as a health protective panacea for diet-related human health problems. However, when criteria for sustainable practice are applied, there are some strong arguments in support of delaying investment in the development and promotion of functional foods to the general public.

Sustainability and technological advances are leading to novel ingredients to emerge from food waste. These sustainable ingredients are finding new applications in foods, nutraceuticals and personal care products. According to the FAO (United Nations), about a third of food produced for human consumption is lost or wasted in the supply chain. With food companies and retailers under pressure to become more efficient and reduce losses, many are re-directing food waste from landfill. A transition is occurring whereby waste previously going to low-end applications – such as animal feed and biogas – is making its way into new ingredients

Novel ingredients are being created from food waste as new technologies improve extraction and processing methods. For instance, the Swiss company FoodSolutionsTeam is using green chemistry to extract active materials from food side streams. Made from carrot pulp, its KaroPRO ingredient has water binding applications in processed foods. The Swiss company has similar food ingredients made from organic linseed, peas and rice. Phytonext is another company using new extraction techniques to produce ingredients from citrus peel and tomato waste.

The EU is also funding research to create new ingredients from food waste. The BIORICE project involves extracting starch from rice waste to make ingredients for functional foods, nutraceuticals and cosmetics. Another project, APROPOS, involves taking proteins from salmon and rapeseed waste for cosmetic applications. A Spanish cosmetics firm plans to use the novel extract as a foundation for a new cream product.

Such ingredients have already made headway in natural personal care products. The Marks & Spencer department store is using resveratrol from grape waste in its Super Grape skin care products. The grape waste is coming from the production of the retailer's own label wines. The French company Caudalie has built its entire range of natural personal care products from grapevine-based and grapeseed ingredients, such as resveratrol and polyphenols. Organic Monitor research finds the brand leads the French natural cosmetics market; its products are now present in over 30 countries.

With the global population projected to reach 9 billion by 2050 and agricultural land becoming increasingly scarce, more investment is expected in creating new ingredients from food waste. A recurring theme in sustainability summits is 'closing loops' whereby waste materials find a new life in fresh applications. Organisations like TerraCycle have been successful in closing the packaging loop for consumer products; it remains to be seen how soon waste from food streams will find better lives.

The major fishing sustainability requirement is an eco-systems approach to fishing and the Marine Stewardship Council's (MSC) (www.msc.org) fishery certification programme and seafood eco-label recognise and reward sustainable fishing. The MSC is a global organisation working with fisheries, seafood companies, scientists, conservation groups and the public to promote the best environmental choice in seafood. Fisheries around the world have been assessed against MSC environmental standards for sustainable fishing and currently at least 63 fisheries are certified in the MSC program with many of these clustered in the Atlantic stretching from Norway to southern Spain. Many supermarket chains now use the blue MSC eco-label and consumers are slowly becoming aware of this label (Gormley 2010b). Consumers eat farmed salmon because they are aware of the health benefits of ω-3 PUFAs. However, feeding vegetable oil rather than marine oil in the feed to salmon and trout leads to fish with a vegetable rather than a marine fat profile. The fish are then 'converted' to marine oil near time of harvest to improve their fat profile. This is deceiving for consumers and highlights the major requirement of finding alternative sources of marine oils containing ω-3 PUFAs. For this reason it may be beneficial to label the EPA/DHA content of farmed oily fish (Gormley, 2012b). Utilisation of seaweeds and microalgae may be an alternative source of ω-3 PUFAs (Soler-Vila et al., 2009), as may single-cell oils (Ward and Singh, 2005). Cronin et al. (1991) indicated big differences between the fat status of wild versus farmed salmon. Despite relatively similar PUFA contents in the flesh oil of both types, the wild salmon had a much higher ratio of ω-3/ω-6 PUFAs 20 than farmed; this is a reflection of the formulated feed used for the latter.

WHAT THE FUTURE HOLDS-CONCLUSIONS

As consumers worldwide become more health conscious, the demand for health-promoting foods and food components is expected to grow. The market for such foods is predicted to be quite large. Before the full market potential can be realized, however, consumers will need to be assured of the safety and efficacy of functional foods. Current and future scientific studies are expected to provide this assurance and to inspire confidence in functional foods in the minds of consumers worldwide.

With a national scope, field-to-table focus, history of accomplishment, and partnering opportunities—as well as national laboratories and capabilities in functional genomics and breeding, agronomics, food science, and human nutrition, the Agricultural Research Service (ARS) is the USDA's principal in-house research agency. ARS is a powerful force in functional foods research and is focusing on developing the following research lines (USDA, 2010):

- Developing lactic probiotics
- Developing prebiotics from crop polysaccharides
- Developing new and improved technologies thatyield health-

promoting foods from cereals, fruit and vegetables, oilseeds, legumes, and dairy products

- Developing methods to reduce food allergens
- Characterizing and evaluating beneficialphytochemicals in blueberries, watermelon,soybeans, crop residues, and food-processing wastes

Additional research is necessary to substantiate the potential health benefits of those foods for which the diet-health relationships are not sufficiently scientifically validated.

Research into functional foods will not advance public health unless the benefits of the foods are effectively communicated to the consumer. Futher trends have been identified based on the needs of the world population as follows:

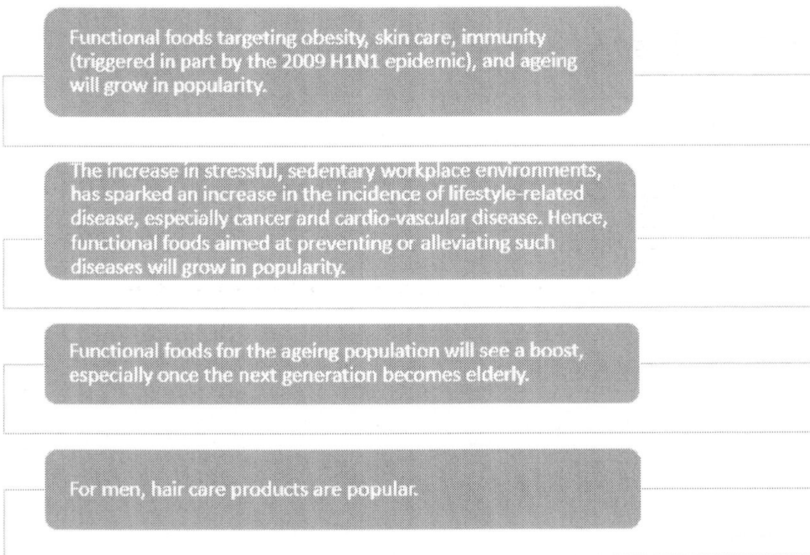

Figure 24: Future trends in functional foods.

The level of consumption of the food that is required to achieve a beneficial effect on health is an important consideration. In particular, it should be possible to achieve the required level of intake of the functional food or ingredient within normal dietary patterns.

Although recent research has shown that growing consumer knowledge about the relationship between diet and health has resulted in a market that should be ready for functional foods, the fortunes of the first products on the market have been mixed. Some are doing very well, while others, despite considerable promotional expenditure and effort being put into explaining them to the consumer, have faced problems on the market, and some have even been withdrawn.

It was mainly the advances in understanding the relationship between nutrition and health that resulted in the development of the concept of functional foods, which means a practical and new approach to achieve optimal health status by promoting the state of well-being and possibly reducing the risk of disease.

Table 23: Major functional food segments

Main Actors	Companies	Functional Products/Ingredients
Multinational food companies with a broad product range.	Nestle, Danone, Unilever Kellogg, Pepsico, Yakult, Danone, Perfetti van Melle	Probiotic Yoghurt/Dahi, Snacks, Energy Drinks, Breakfast Cereals, Confectionary
Pharmaceutical and/or dietary products producing companies.	GSK, Amway, Ranbaxy	Malted Food, Supplements Fortified products.
National "category leaders"	Amul, Dabur, ITC, Britannia, Parle	Dairy Products, Fruits & Vegetable juices, Biscuits
Small and medium-sized Companies (SMEs) of the Food industry	Heritage Foods, Ruchi Soya, Mother Dairy	Soya Milk, Dairy, Oils
Retail companies	Reliance Wellness, Apollo Pharmacy, Patanjali	Sweetners, Cereals, Energy Drinks, Ayurvedic & Herbal Products
Supplier of "functional Ingredients"	Chr Hansen, Orana Danisco	Cultures, Enzymes, Phytonutrients, Natural Colors

Adapted from Menrad, 2003

Functional foods are found virtually in all food categories;however, products are not homogeneously scattered over all segments of the growing market. The development and commerce of these products is rather complex, expensive and risky, as special requirements should be answered. Besides potential technological obstacles, legislative aspects, as well as consumer demands need to be taken into consideration when developing functional food. In particular, consumer acceptance has been recognized as a key factor to successfully negotiate market opportunities.

Another potential of functional foods is to have edible vaccines. Currently, vaccine treatment for all is far from reality. It is also expensive, and requires refrigeration and a skilled person to give the injection. Health expenditure in many countries is declining (in part to things like flawed structural adjustment impositions by the World Bank and IMF). Even needles are even hard to come by in some places.

As a result, edible vaccines sound like a promising alternative. However, this is another technology which critics point a lot of holes in, as, described in an article titled, "Eat Up Your Vaccines", (December 2000) from Genetic Resources Action Internatinal (GRAIN). For example, the article points out concerns on things like the ones detailed on the table below.

Table 24: Identified concerns with edible vaccines

Being able to determined and assure the right dosage is taken. As an example, the article points out that "Charles Arntzen, one of the pioneers of edible vaccines, acknowledges the challenge of assessing how much an infant, in particular, ingests. "A baby may eat a bite and not want any more, may spit up half of it, or eat it all and throw it up later," he concedes."
Growing the foods to deliver these have additional problems, due to the need of sufficient amounts of proteins.
Limitations of edible vaccines; that "[a]nother much-hyped advantage ignores the fact that if they could be given orally, today's vaccines already would be. Few vaccines are absorbed well from the gut because they are too big to cross the gut wall easily and/or are broken down by the gut enzymes. Edible vaccines would be subject to the same limitations as any other oral drugs."
There will still be considerable costs, not real reductions.
Being a new idea, it is not clear if it will work everywhere, for economic and biological reasons.

Environmental risks; the fear of horizontal gene transfer (the transfer of genes from one species to another, unrelated, species) will have numerous unpredictable and dangerous effects. "Antibiotics and traditional vaccines already contribute to horizontal gene transfer. Recombinant vaccines, like those that would be used in edible vaccines, would exacerbate such transfer. This is a serious concern for the release of any genetically manipulated organism, but particularly worrisome in the case of vaccines, because of their disease-causing potential."

The Challenge of Unlocking the Food/Health Equation

A number of challenges exist for the vitality of the functional foods market in the years ahead. First, there are a number of scientific challenges for those attempting to unlock the food/health "equation." How do these food components act to modify disease processes and for whom are these benefits most pronounced? The food/health equation includes a complex set of interactions that must be better understood so that new food product efficacy can be appropriately characterized. The factors contributing to disparities in individual health responses to foods need to be clarified. These include genetic variation, gender differences, individual differences in metabolism, and how bioactive phytochemicals alter the routine functions of traditional nutrients. This challenge is underscored by the emergence of the field of nutrigenomics, in which investigators worldwide are studying how food constituents acting as "dietary signals" can modulate multiple gene-expression patterns in our cells and tissues, and ultimately reduce disease risk (Kaput and Rodriguez 2004). Another scientific challenge involves understanding our complex behaviors associated with food consumption. For example, the sheer magnitude of sensory dynamics is illustrated by the fact that we may consume over 10,000 different food chemicals in our diet which can interact with thousands of chemosensor receptors around our bodies (Le Coutre 2003). How is this information coded in the human brain and how does it impact eating behaviors? Related to this issue —and perhaps one of the greatest challenges confronting those attempting to unlock the food/health equation—is gaining a meaningful understanding of the contributors to the international prevalence of obesity.

Advancing Functional Foods Research

Extensive research has shown that the bioavailability of nutrients within a given food is governed by the composition and microstructure of the food, commonly referred to as the food matrix (Hwang et al., 2000). More specifically, numerous studies investigating the development of functional foods have shown that the interactions between nutrients and non-nutrients within the food matrix can be synergistic, additive, or neutralizing in nature (Jeffery, 2005). For example, vitamin C regenerates vitamin E and enhances the antioxidant effect of carotenoid compounds (Yeum et al., 2004). Likewise, in vitro studies have shown that flavonoids, a subcategory of phenolic phytochemicals, behave synergistically with vitamin E to prevent low-density lipoprotein oxidation (Hwang et al., 2000). Thus, given the diversity of interactions that exist among nutrient and non-nutrient compounds, the interactive effects of these compounds must be fully elucidated in order to develop functional foods with the greatest potential to synergistically impact human health.

The industry perspective

Food is a business of brands in a modern trading market. It is the brand that establishes trusted, known and appealing images to its audience. Brands can be extended by value-adding with the Functional Foods concept, however, this needs to be done strategically. If positioned incorrectly, this move can undermine existing brands. The aim of the functional food industry is to differentiate and value add to food industry initiatives by integrating knowledge from market trends, science/ technology, regulation and industry/company capability. The functional food industry helps to develop solutions to diminish risk from the science and regulatory interface. It facilitates the integration of expertise and knowledge from concept to commercialisation by working in partnership with industry, acknowledging its experience in the field. To ensure consumer acceptance, companies need to address issues associated with the taste, texture, smell, safety and stability of their products. In addition, they also need to attend to packaging and distribution issues. With functional foods, the ability to deliver bioactive components in the required amounts and with demonstrated bioactivity may require advanced expertise such as microencapsulation and extraction technologies. These technological challenges may be addressed in-house or may require the input of specialist providers.

As for trends in the world of functional foods, the rise of single male households and an ageing population were the top two of 10 key food and drink trends that will dominate the food industry this year offering opportunities for manufacturers, according to the influential Chicago-based Institute of Food Technologists (IFT). Below are the top 10 food trends as identified by IFT:

1. **Men in the Aisles:** Single male households are on the rise and men now account for 43 percent of primary shopping and 46 percent help with food preparation (FMI 2015). Claims related to sodium content are very important to 56 percent of male shoppers, followed by claims related to sugar content, calories, total fat, trans fat, cholesterol, saturated fat and protein (Packaged Facts).

2. **Gray Matters:** By 2019, half of the U.S. population will be over 50 years old and will account for more than half of all consumer product spending (IRI 2014). Adults aged 55-64 are the most likely to be trying to lose weight; those aged 65-plus are most likely to be trying to maintain weight (Packaged Facts 2015).

3. **Chemical Consciousness:** Clean claims have more appeal than generally healthy descriptors. Preservative-free is second only to no trans-fat/no partially hydrogenated oils as the most important consumer food claim (Mintel 2015). Eight in 10 adults define preservative-free foods/beverages as healthy, and 78 percent say no artificial ingredients translates to healthy. Other attributes associated with healthfulness include antibiotic- or hormone free, natural, unprocessed, organic, not genetically modified, clean and real (Technomic 2014).

4. **Personalized Nutrition Plans:** In 2015, 30 percent of food shoppers participated in at least one specialized eating plan (FMI

2014). Millennials were the most likely to adopt a specialized eating approach as well as the most likely to try more than one plan and tended to favor vegetarian, juice cleanse/detox, Paleo, or vegan styles of eating. Older generations were most likely to go lactose-free (FMI 2015).

5. **A New Natural Hierarchy:** Minimally processed foods/ natural ingredients were cited for the first time among the "hot" culinary trends for 2016 identified by the American Culinary Federation (NRA 2015). One quarter of consumers say that their food habits characterize them as living a minimally processed lifestyle and 30 percent are making a strong effort to eat more minimally processed foods (NBJ 2015, MSI 2014). Although half of consumers are still unsure of the definition of GMO, after hearing a brief definition 45 percent would avoid GMOs if they could (FMI 2015).

6. **Lifestyle Enhancers:** Consumers, especially women and young adults, are increasingly concerned about conditions that impact their ability to function day by day. Food/beverages that provide energy throughout the day appeal to 30 percent of adults, 26 percent are looking for more mental energy, 14 percent morning energy and 7 percent an energy boost later in the day (HealthFocus 2015).

7. **Definitely Deficient:** According to the *2015 Dietary Guidelines for Americans,* Americans are not consuming enough of some nutrients. Sixty-four percent of adults regularly looked for foods/ beverages with added vitamins or minerals in 2015. Although this is a 9 percent drop over the past 5 years, 82 percent feel fortified foods are a convenient way to get their nutrients and that they're extremely important for children (Hartman 2015, MSI 2014).

8. **Food MD:** High cholesterol, followed by weight loss, high blood pressure, weight management, digestive health, improving heart health, energy, immunity and joint/bone health are the top health concerns motivating consumers to seek out functional foods/ beverages. (Packaged Facts, 2015).

9. **Wholly Functional:** One-third of consumers strongly agree that naturally occurring nutrients in foods are better for them than added nutrients, 36 percent agree somewhat and 30 percent of adults describe themselves as living a "whole foods" lifestyle (Packaged Facts 2015, NBJ 2015).

10. **Healthy Convenience:** One in five adults eats on the run, grabbing food whenever they can (Wyatt 2015). Fifty-five percent of households have at least one brown-bagger and 20 percent bring beverages from home. Snacks account for nearly half of all meal occasions and 16 percent of all eating occasions are immediate consumption meaning food eaten within 60 minutes of purchase (Hartman 2014).

Opportunities for further development of functional foods

The best opportunities for Functional Foods can be seen as those where the market is ready; the underpinning science to substantiate benefits is available; the regulatory requirements are achievable; and the company has the technical support and ability to develop and commercialise the food product. For the major export industries such as meat, grains, dairy and wine, a stronger competitive advantage may be achieved by increasing export of value referenced commodities and/or value added processed foods. This would be in addition to their unique benefits already offered by Australian foods of combined product quality, food safety and environmental friendliness. Rapid urbanisation, health concerns and demand for convenience present many opportunities for Functional Foods which combine the need for both health and convenience. Specific markets for functional food opportunities are identified in this paper, particularly in Asia. The most developed areas of supportive science relate to lifestyle diseases (obesity, heart disease and diabetes) and bone health. Further scientific development is required in other areas of functionality such as performance, gut health, immunity and cancer prevention. The challenge is to develop effective synergies between science, technology and food product development for the benefit of the consumer.

An analysis of market, science and regulatory factors reflects opportunities for Functional Foods for healthy aging, appetite control and kids' nutrition. Demand for foods for appetite control will increase with growing awareness of oesity as a problem. Currently the science and regulatory environments are targeting foods that may help reduce energy intake. Interest in kid's nutrition may increase as the number of children per family unit decreases. Opportunities for Functional Foods will need to be considered in view of ethical marketing and regulatory standards. Functional Foods present real opportunities in food innovation. This paper has identified indicative areas for further exploration but needs to be blended with the brand owners' knowledge and experience to drive the food innovation process.

REFERENCES

1. Aagaard, K., Riehle, K., Ma, J., Segata, N., Mistretta, T.-A., Coarfa, C., Raza, S., Rosenbaum, S., Van den Veyver, I., Milosavlijevic, A., Gevers, D., Huttenhower, C., Petrosino, J. & Versalovic, J. (2012). A metagenomic approach to characterization of the vaginal microbiome signature in pregnancy. 7(6), e36466. doi: 10.1371/journal.pone.0036466
2. AbuMweis S, Barake R, Jones PJH. Plant sterols/stanols as cholesterol lowering agents: A meta-analysis of randomized controlled trials. Food & Nutrition Research. 2008;52. doi: 10.3402/fnr.v52i0.1811.
3. Acute fish oil and soy isoflavone supplementation increase postprandial serum (n-3) polyunsaturated fatty acids and isoflavones but do not affect triacylglycerols or biomarkers of oxidative stress in overweight and obese hypertriglyceridemic men. Journal of Nutrition, 139(6), 1128e1134.

4. Adlercreutz, H., Fotsis, T., Heikkinen, R., Dwyer, JT, Woods, M., Goldin, BR, and Gorbach, SL (1982). Excretion of the lignans enterolactone and enterodiol and of equol in omnivorous and vegetarian postmenopausal women and in women with breast cancer. Lancet ii, p.1295-1299. 1062-1076.

5. Age-Related Eye Disease Study 2 Research Group, Chew EY, Clemons TE, et al. Secondary analyses of the effects of lutein/zeaxanthin on age-related macular degeneration progression: AREDS2 report No. 3. JAMA Ophthalmol. 2014;132(2):142-149. (PubMed)

6. Age-Related Eye Disease Study 2 Research Group. Lutein + zeaxanthin and omega-3 fatty acids for age-related macular degeneration: the Age-Related Eye Disease Study 2 (AREDS2) randomized clinical trial. JAMA. 2013;309(19):2005-2015. (PubMed)

7. Age-Related Eye Disease Study Research Group. A randomized, placebo-controlled, clinical trial of high-dose supplementation with vitamins C and E, β-carotene, and zinc for age-related macular degeneration and vision loss: AREDS report no. 8. Arch Ophthalmol. 2001;119(10):1417-1436. (PubMed)

8. Agriculture and Agri-Food Canada. What are Functional Foods and Nutraceuticals; 2009 [Accessed 22.02.17]. Available from: http://www4.agr.gc.ca/AAFC-AAC/displayafficher.do?id=1171305207040&lang=eng.

9. Ahn, I.-S., Do, M.-S., Kim, S.-O., Jung, H.-S., Kim, Y.-I., Kim, H.-J., et al. (2006). Antiobesity effect of Kochujang (Korean fermented red pepper paste) extract in 3T3-L1 adipocytes. Journal of Medicinal Food, 9(1), 15e21.

10. Ahrens EH Jr, Boucher CA. The composition of a simulated American diet. J Am Diet Assoc. 1978;73:613–620.

11. Alasalvar, C. and Taylor, Y. (eds). 2002. Seafoods – Quality Technology and Nutraceutical Applications. Springer-Verlag, Heidelberg, Germany

12. Alasalvar, C. and Taylor, Y. (eds). 2002. Seafoods – Quality Technology and Nutraceutical Applications. Springer-Verlag, Heidelberg, Germany

13. Albertazzi, P., Pansini, F., Bonaccorsi G., Zanotti, L., Forini, E., and De Aloysio, D. (1998). The effect of dietary soy supplementation on hot flushes. Obstetrics and Gynaecology 91, p. 6-11.

14. Ali, A. A., Velasquez, M. T., Hansen, C. T., Mohamed, A. I., & Bhathena, S. J. (2005). Modulation of carbohydrate metabolism and peptide hormones by soybean isoflavones and probiotics in obesity and diabetes. Journal of Nutritional Biochemistry, 16(11), 693e699.

15. Alles MS, de Roos NM, Bakx JC, van de Lisdonk E, Zock PL, Hautvast GA. Consumption of fructooligosaccharides does not favorably affect blood glucose and serum lipid concentrations in patients with type 2 diabetes. Am J Clin Nutr 1999;69:64–9.

16. Alles MS, Hartemink R, Meyboom S, et al. Effect of transgalactooligosaccharides on the composition of the human intestinal microflora and on putative risk markers for colon cancer. Am J Clin Nutr 1999;69:980–91.

17. Allman, MA., Pena, MM., and Pang, D. (1995). Supplementation with flaxseed oil versus sunflowerseed oil in healthy young men consuming a low fat diet: Effects on platelet composition and function. European Journal of Clinical Nutrition 49, p.169-178.

18. American Dietetic Association (1999) Position of the American Dietetic Association: functional foods. J. Am. Diet. Assoc. 99:1278-1285. MedlineGoogle Scholar

19. American Dietetic Association: Position of the American Dietetic Association: functional foods. Journal of the American Dietetic Association 1999, 99: 1278-1285. Functional Foods in Health and Disease 2015; 5(6):209-223 Page 222 of 223

20. American Dietetic Association: Position of the American Dietetic Association: Functional Foods. J Am Diet Assoc. 2009;109:735-746.

21. Anand, P., Kunnumakkara, A. B., Newman, R. A., & Aggarwal, B. B. (2007). Bioavailability of curcumin: Problems and promises. Molecular Pharmaceutics, 4(6), 807e818.

22. Anderson, JW., Johnstone, BM. and Cook-Newell, ME. (1995). Meta-analysis of the effects of soy protein in-take on serum lipids. New England Journal of Medicine 333, p. 276- 282.

23. Anderson, R. A. (2008). Chromium and polyphenols from cinnamon improve insulin sensitivity. Proceedings of Nutrition Society, 67(1), 48e53.

24. Andre, C. M., Oufir, M., Guignard, C., Hoffmann, L., Hausman, J.-F., Evers, D., et al. (2007). Antioxidant profiling of native Andean potato tubers (Solanum tuberosum L.) reveals cultivars with high levels of B-carotene, a-tocopherol, chlorogenic acid, and petanin. Journal of Agricultural and Food Chemistry, 55(26), 10839e10849.

25. Andujar, I., Recio, M. C., Giner, R. M., & Rios, J. L. (2012). Cocoa polyphenols and their potential benefits for human health. Oxidative Medicine and Cell Longevity, 906252.

26. Anese M & Nicoli MC (2003) Antioxidant properties of ready-to-drink coffee brews. J Agric Food Chem 51, 942–946.

27. Ansari MS, Gupta NP. A comparison of lycopene and orchidectomy vs orchidectomy alone in the management of advanced prostate cancer. BJU Int. 2003;92(4):375-378; discussion 378. (PubMed)

28. Aprikian, O., Busserolles, J., Manach, C., Mazur, A., Morand, C., Davicco, M.-J., et al. (2002). Lyophilized apple counteracts the development of hypercholesterolemia, oxidative stress, and renal dysfunction in obese zucker rats. Journal of Nutrition, 132(7), 1969e1976.

29. Arai, S. (1996) Studies on functional foods in Japan—state of the art. Biosci. Biotechnol. Biochem. 60:9-15.

30. Arai S (1996). Studies on functional foods in Japan. Bioscience Biotechnol Biochem 60, 9–15.

31. Ares G, Gambaro A. Influence of gender, age and motives underlying food choices and perceived healthiness and willingness to try functional foods. Appetite. 2007;49:148-58

32. Ares G, Gimenez A, Gambaro A. Influence of nutritional knowledge on perceived healthiness and willingness to try functional foods. Appetite. 2008;51:663-668

33. Ares, G., and Gámbaro, A. (2007). Influence of gender, age and motives underlying food choice on perceived healthiness and willingness to try functional foods. Appetite, 49, 148–158.

34. Arya, SS., Salve, AR., and Chauhan, S. (2016). Peanuts as functional food: a review. J Food Sci Technology 53(1), pp. 31–41.

35. Arvanitoyannis IS, Van Houwelingen-Koukaliaroglou M (2005). Functional foods: a survey of health claims, pros and cons, and current legislation. Crit Rev Food Sci Nutr 45, 385–404.

36. Asdaq, S. M. B., & Inamdar, M. N. (2011). Pharmacodynamic and pharmacokinetic interactions of propranolol with garlic (Allium sativum) in rats. Evidence-based Complementary and Alternative Medicine, 24042. http://dx.doi.org/10.1093/ecam/neq076

37. Ashraf, R., Khan, R. A., & Ashraf, I. (2011). Garlic (Allium sativum) supplementation with standard antidiabetic agent provides better diabetic control in type 2 diabetes patients. Pakistan Journal of Pharmaceutical Science, 24(4)

38. Aune D, Chan DS, Vieira AR, et al. (2012). Dietary compared with blood concentrations of carotenoids and breast cancer risk: a systematic review and meta-analysis of prospective studies. Am J Clin Nutr. 96(2):356-373

39. Australia New Zealand Food Authority ANZFA (2001). Phytosterol esters derived from vegetable oils: Final Risk Analysis Report Application A410, 30/5/01. Available from: http://www.anzfa.gov.au/documents/A410_IR(FULL).pdf Accessed 19.02.17

40. Aviram M, Dornfeld L, Rosenblat M, Volkova N, Kaplan M, et al. (2000) Pomegranate juice consumption reduces oxidative stress, atherogenic modifications to LDL, and platelet aggregation: studies in humans and in atherosclerotic apolipoprotein E-deficient mice. Am J Clin Nutr 71:

41. Avorn, J., Monane, M., Gurwitz, JH., Glynn, RJ., Choodnovskiy, I., and Lipsitz, LA. (1994). Reduction of bacteriuria and pyuria after ingestion of cranberry juice -- A reply. Journal of the American Medical Association 272, p. 589-590.

42. Awad AB, Roy R, Fink CS (2003). Beta-sitosterol, a plant sterol, induces apoptosis and activates key caspases in MDA-MB-231 human breast cancer cells. Oncol Rep. 10:497–500

43. Ayyagari, R., Grover V. and Purvis R. (2011). Technostress: Technological Antecendents and Implications. MIS Quarterly, 35(4), 831-858

44. Azad, M.B., Konya, T., Maughan, H., Guttman, D.S., Field, C.J., Chari, R.S., Sears, M.R., Becker, A.B., Scott, J.A. & Kozyrskyj, A.L. (2013). Gut microbiota of healthy Canadian infants: profiles by mode of delivery and infant diet at 4 months. Canadian Medical Association Journal, 185(5), 385-394. doi: 10.1503/cmaj.121189
45. Bäckhed F, Ding H, Wang T, Hooper LV, Koh GY, et al. (2004) The gut microbiota as an environmental factor that regulates fat storage. Proceedings of the National Academy of Sciences USA 101: 15718-15723.
46. Bäckhed F, Roswall J, Peng Y, Feng Q, Jia H, et al. (2015) Dynamics and Stabilization of the Human Gut Microbiome during the First Year of Life. Cell Host Microbe 17: 690-703.
47. Bagchi D (ed). (2008). Neutraceutical and Functional Food Regulations. Elsevier: New York.
48. Bahri, N., Gusko, A., Hamm, M., Kasper, H., Klor, H.U., Neuberger, D. and Singer, P. (2002). Significance and recommended dietary intake of long-chain omega-3 fatty acids- A consensus statement of the omega-3 working group. Ernahrungs-Umschau, 49(3), 94-98.
49. Bakker MF, Peeters PH, Klaasen VM, et al. (2016). Plasma carotenoids, vitamin C, tocopherols, and retinol and the risk of breast cancer in the European Prospective Investigation into Cancer and Nutrition cohort. Am J Clin Nutrition 103(2):454-464.
50. Balasubramanian S, Cole C. Consumers' Search and Use of Nutrition Information: The Challenge and Promise of the Nutrition Labeling and Education Act. J Marketing. 2002;66:112-127.
51. Banerjee A. and Banerjee B. (2000). Effective Retail Promotion Management: Use of Point of Sales Information Resources. Vikalpa, 25(4), 52-59.
52. Baret, P., Septembre-Malaterre, A., Rigoulet, M., Lefebvre, D'Hellencourt, C., Priault, M., et al. (2013). Dietary polyphenols preconditioning protects 3T3-L1 preadipocytes from mitochondrial alterations induced by oxidative stress. International Journal of Biochemistry and Cell Biology, 45(1), 167e174.
53. Barker, K. (2006). Fast food greases India's way to fat. Available at: http://articles.chicagotribune.com [Accessed 22.02.17]

54. Barile D, Rastall RA (2013) Human milk and related oligosaccharides as prebiotics. Current Opinions in Biotechnology 24: 214-219.

55. Bayele HK., Balesaria S., and Srai SK. (2015). Phytoestrogens modulate hepcidin expression by Nrf2: Implications for dietary control of iron absorption. Free Radic Biol Med. 89:1192-202.

56. Beatty S, Murray IJ, Henson DB, Carden D, Koh H, Boulton ME (2001). Macular pigment and risk for age-related macular degeneration in subjects from a Northern European population. Invest Ophthalmol Vis Sci. 42(2):439-446.

57. Bechtel, P.J. (ed.). 2002. Advances in Seafood By-Products: 2002 Conference Proceedings. Alaska Sea Grant Program, University of Alaska, Fairbanks, Alaska, USA

58. Bech-Larsen T, Grunet K. (2003). The perceived healthiness of functional foods A conjoint study of Danish, Finnish and American consumers'perception of functional foods. Appetite. 40:9-14

59. Bech-Larsen, T. and Scholderer, J. (2007). Functional foods in Europe: Consumer research, market experiences and regulatory aspects. Trends in Food Science & Technology, 18, 231–234

60. Becker M, Staab D & von Bergmann K (1992) Treatment of severe familial hypercholesterolemia in children: effect of sitosterol and bezafibrate. Pediatrics 89, 138–142.

61. Becker M, Staab D & Von Bergmann K (1993) Treatment of severe familiar hypercholesterolemia in childhood with sitosterol and sitostanol. Journal of Pediatrics 122, 292–296.

62. Belasco W (1989) Appetite For Change: How The Counter Culture Took on The Food Industry, 1966-1988. Ithaca: Cornell University Press.

63. Belury, M.A. (2002) Dietary conjugated linoleic acid in health. Annual Review of Nutrition, 22, 505-531.

64. Benbrook CM, Butler G, Latif MA, Leifert C, Davis DR (2013). Organic production enhances milk nutritional quality by shifting fatty acid composition: a United States-wide, 18 month study. PloS One 8:e82429.

65. Bercik P, Denou E, Collins J, Jackson W, Lu J, et al. (2011) The intestinal microbiota affect central levels of brain-derived neurotropic factor and behavior in mice. Gastroent 141: 599-609.

66. Berge, R. 2005. Fish protein hydrolysates. Patent Number WO 2005002605 A1.

67. Berger, A., Jones, PJH., and Abumweis SS (2004). Plant sterols: factors affecting their efficacy and safety as functional food ingredients. Lipids Health Dis. 3: 5.

68. Best MM, Duncan CH, van Loon EJ & Wathen JD (1955) The effect of sitosterol on serum lipids. American Journal of Medicine 15, 61–70

69. Bhadoria P.B.S and Mahapatra S.C (2011). Prospects, Technological Aspects and Limitations of Probiotics – A Worldwide Review. European Journal of Food Research & Review, 1(2), 23-42.

70. Bierenbaum, ML, Reichstein, R. and Watkins, TR (1993). Reducing atherogenic risk in hyperlipemic humans with flax seed supplementation: A preliminary report. Journal of the American College of Nutrition 12, p.501-504.

71. Biltekoff C (2010) Consumer response: 3e paradoxes of food and health. Annals of the New York Academy of Science 1190 174–178.

72. Bioavailability and Bioequivalence Requirements: General Definitions, 21 Code of Federal Regulations Sub-part A 320.1 (2012)

73. Bjelakovic G, Nikolova D, Gluud LL, Simonetti RG, Gluud C (2007). Mortality in randomized trials of antioxidant supplements for primary and secondary prevention. JAMA. 297(8):842-857

74. Blatherwick, NR. (1914). The specific role of foods in relation to the composition of the urine. Archives of Internal Medicine 14, p. 409-450.

75. Block G (1993). Micronutrients and cancer: time for actions? Journal of the National Cancer Institute 85: 846-848.

76. Block, E. (1992b). The organosulfur chemistry of the genus Allium--Implications for the organic chemistry of sulfur. Angewandte Chemie International Edition in English 31, p. 1135-1178.

77. Block, G., Patterson, B. and Subar, A. (1992). Fruit, vegetables, and cancer prevention: A review of the epidemiological evidence. Nutrition and Cancer 18, p. 1-29.

78. Block, R.C. and Pearson, T.A. (2006). The cardiovascular implication of omega-3 fatty acids. Cardiology Journal, 13(7), 557-569.

79. Boath, A. S., Grussu, D., Stewart, D., & McDougall, G. J. (2012). Berry polyphenols inhibit digestive enzymes: a source of potential health benefits? Food Digestion, 3(3-Jan), 1e7.

80. Bone RA, Landrum JT, Mayne ST, Gomez CM, Tibor SE, Twaroska EE (2001). Macular pigment in donor eyes with and without AMD: a case-control study. Invest Ophthalmol Vis Sci. 42(1):235-240.

81. Bosner MS, Lange LG, Stenson WF, Ostlund Jr RE (1999). Percent cholesterol absorption in normal women and men quantified with dual stable isotopic tracers and negative ion mass spectrometry. J Lipid Res 40:302–308.

82. Botelho, F. V., Eneas, L. R., Cesar, G. C., Bizzotto, C. S., Tavares, E., Oliveira, F. A., et al. (2004). Effects of eggplant (Solanum melongena) on the atherogenesis and oxidative stress in LDL receptor knock out mice (LDLR-/-). Food and Chemical Toxicology, 42(8), 1259e1267.

83. Bots ML, Grobbee DE (2002). Intima media thickness as a surrogate marker for generalised atherosclerosis. Cardiovasc Drugs Ther. 16(4):341-351. (PubMed)

84. Bouhnik Y, Flourie B, D'Agay-Abensour L, et al (1997). Administration of transgalacto-oligosaccharides increases fecal bifidobacteria and modifies colonic fermentation metabolism in healthy humans. J Nutr 127:444–8

85. Bouic PJ (2001). The role of phytosterols and phytosterolins in immune modulation: a review of the past 10 years. Curr Opin Clin Nutr Metab Care. 4:471–475.

86. Bourassa MW, Alim I, Bultman SJ, Ratan RR (2016) Butyrate, neuroepigenetics and the gut microbiome: Can a high fiber diet improve brain health? Neurosci Lett 625: 56-63.

87. Boyer J, Liu RH (2004). Apple phytochemicals and their health benefits. Nutr Journal 3(5):5-9

88. Bracesco, N., Sanchez, A. G., Contreras, V., Menini, T., & Gugliucci, A. (2011). Recent advances on Ilex paraguariensis research: minireview. Journal of Ethnopharmacology, 136(3), 378e384.

89. Braida, M. and Gormley, T.R. (2008). Fish ready-meals with nutraceuticals. Food Science and Technology, 22 (3), 26-28.

90. Broadhurst, C. L., Polansky, M. M., & Anderson, R. A. (2000). Insulinlike biological activity of culinary and medicinal plant aqueous extracts in-vitro. Journal of Agricultural and Food Chemistry, 48(3), 849e852.

91. Brower V (1998). Nutraceuticals: poised for a healthy slice of the healthcare market? Nat. Biotechnol.16:728–731.

92. Brown B, Zhao XQ, Chait A, et al. (2001). Simvastatin and niacin, antioxidant vitamins, or the combination for the prevention of coronary disease. N Engl J Med. 345(22):1583-1592.

93. Brown, A. L., Lane, J., Coverly, J., Stocks, J., Jackson, S., Stephen, A., et al. (2009). Effects of dietary supplementation with the green tea polyphenol epigallocatechin-3-gallate on insulin resistance and associated metabolic risk factors: randomized controlled trial. British Journal of Nutrition, 101(6), 886e894.

94. Brown, A. L., Lane, J., Holyoak, C., Nicol, B., Mayes, A. E., & Dadd, T. (2011). Health effects of green tea catechins in overweight and obese men: a randomised controlled cross-over trial. British Journal of Nutrition, 106(12), 1880e1889.

95. Brown, EN. (2012). Plant Sterols: The Next Heart Healthy Food Fad [online]. Available from: http://www.blisstree.com/2012/02/17/food/nutrition/plant-sterols-heart-health-fortified-foods-280/. Accessed 20.02.17

96. Bueno, A. A., Oyama, L. M., De Oliveira, C., Pisani, L. P., Ribeiro, E. B., Silveira, V. L. F., et al. (2008). Effects of different fatty acids and dietary lipids on adiponectin gene expression in 3T3-L1 cells and C57BL/6J mice adipose tissue. Pflugers Archiv e European Journal of Physiology, 455(4).

97. Buijsse B, Feskens EJ, Kwape L, Kok FJ, Kromhout D (2008). Both alpha- and β-carotene, but not tocopherols and vitamin C, are inversely related to 15-year cardiovascular mortality in Dutch elderly men. J Nutrition 138(2):344-350

98. Burr, M.L., Fehily, A.M. and Gilbert, J.F. (1989). Effects of changes in fat, fish and fibre intakes on death and myocardial reinfarction: diet and reinfarction trial (DART). Lancet 2, 757-761

99. Burton-Freeman, B., Linares, A., Hyson, D., & Kappagoda, T. (2010). Strawberry modulates LDL oxidation and postprandial

lipemia in response to high-fat meal in overweight hyperlipidemic men and women. Journal of the American College of Nutrition, 29(1), 46e54.

100. Byrne CS, Chambers ES, Alhabeeb H, Chhina N, Morrison DJ, et al. (2016) Increased colonic propionate reduces anticipatory reward responses in the human striatum to high-energy foods. He Am J Clin Nutr 104: 5-14.

101. Cabrera, C., Artacho, R., & Gimenez, R. (2006). Beneficial effects of green tea e a review. Journal of the American College of Nutrition, 25(2), 79e99.

102. Cannon CP, Braunwald E, McCabe CH, Rader DJ, Rouleau JL, Belder R, Joyal SV, Hill KA, Pfeffer MA, Skene AM (2004). Comparison of Intensive and Moderate Lipid Lowering with Statins after Acute Coronary Syndromes. N Engl J Med. 350:15.

103. Cao, H., Hininger-Favier, I., Kelly, M. A., Benaraba, R., Dawson, H. D., Coves, S., et al. (2007). Green tea polyphenol extract regulates the expression of genes involved in glucose uptake and insulin signaling in rats fed a high fructose diet. Journal of Agricultural and Food Chemistry, 55(15), 6372e6378.

104. Cao, H., Polansky, M. M., & Anderson, R. A. (2007). Cinnamon extract and polyphenols affect the expression of tristetraprolin, insulin receptor, and glucose transporter 4 in mouse 3T3-L1 adipocytes. Archives of Biochemistry and Biophysics, 459(2), 214e222.

105. Caporaso, J.G., Lauber, C.J., Costello, E.K., Berg-Lyons, D., Gonzalez, A., Stombaugh, J., Knights, D., Gajer, P., Ravel, J., Fierer, N., Gordon, J.I. & Knight, R. (2011). Moving pictures of the human microbiome. Genome Biology, 12, R50. doi: 10.1186/gb-2011-12-5-r50

106. Cara J. Westmark (2014). Definition of Functional Food. Healthy, Functional, and Medical Foods. Similarities and Differences between these Categories. Bioactive Food Compounds. Introduction to Functional Food Science: Textbook. 2nd ed. Richardson, TX: Functional Food Center

107. Carr TP, Cornelison RM, Illston BJ, Stuefer-Powell CL, Gallaher DD (2002). Plant sterols alter bile acid metabolism and reduce

cholesterol absorption in hamsters fed a beef-based diet. Nutrition Research 22:745–754

108. Casal S, Oliveira MB, Alves MR & Ferreira MA (2000) Discriminate analysis of roasted coffee varieties for trigonelline, nicotinic acid, and caffeine content. J Agric Food Chem 48, 3420–3424
109. Casas-Agustench, P., Serra, M., Perez-Heras, A., Cofan, M., Pinto, X., Trautwein, E. A., et al. (2012). Effects of plant sterol esters in skimmed milk and vegetable-fat-enriched milk on serum lipids and non-cholesterol sterols in hypercholesterolaemic subjects: a randomised, placebo-controlled, crossover study. British Journal of Nutrition, 107(12), 1766e1775.
110. Castro-Gonzalez, M.I., Ojeda, V.A., Montano, B.S., Ledesma, C.E. and Perez-Gil, R.F. (2007). n-3 fatty acid evaluation in eighteen Mexican marine fishes as functional food. Archivos LatinoAmericanos de Nutricion, 57(1), 85-93
111. Chaturvedi, P. (2012). Antidiabetic potentials of momordica charantia: multiple mechanisms behind the effects. Journal of Medicinal Food, 15(2), 101e107.
112. Chen, C.-F., Li, Y.-D., & Xu, Z. (2010). Chem principles and bioactivities of blueberry. Yaoxue Xuebao, 45(4), 422e429.
113. Cheng, D. M., Kuhn, P., Poulev, A., Rojo, L. E., Lila, M. A., & Raskin, I. (2012). In vivo and in vitro antidiabetic effects of aqueous cinnamon extract and cinnamon polyphenol-enhanced food matrix. Food Chemistry, 135(4), 2994e3002.
114. Cherniack, E. P. (2011). Polyphenols: Planting the seeds of treatment for the metabolic syndrome. Nutrition, 27(6), 617e623.
115. Chew EY, Clemons TE, Agron E, et al. (2015). Effect of Omega-3 Fatty Acids, Lutein/Zeaxanthin, or Other Nutrient Supplementation on Cognitive Function: The AREDS2 Randomized Clinical Trial. JAMA. 314(8):791-801.
116. Cho E, Seddon JM, Rosner B, Willett WC, Hankinson SE (2004). Prospective study of intake of fruits, vegetables, vitamins, and carotenoids and risk of age-related maculopathy. Arch Ophthalmol. 122(6):883-892.

117. Choi, E.-M. (2007). Suppressive effects of potato (Solanum tuberlosum) on type II collagen-induced arthritis in DBA/1J mice. Food Science and Biotechnology, 16(1), 43e48.

118. Choi, K.-M., Lee, Y.-S., Shin, D.-M., Lee, S., Yoo, K.-S., Lee, M. K., et al. (2013). Green tomato extract attenuates high-fat-dietinduced obesity through activation of the AMPK pathway in C57BL/6 mice. Journal of Nutritional Biochemistry, 24(1), 335e342.

119. Chonpathompikunlert, P., Wattanathorn, J., & Muchimapura, S. (2010). Piperine, the main alkaloid of Thai black pepper, protects against neurodegeneration and cognitive impairment in animal model of cognitive deficit like condition of Alzheimer's disease. Food Chemistry and Toxicology, 48(3), 798e802.

120. Choo, H., Chung, J. and Pysarchik, D.T. (2004). Antecedents to new food product purchasing behavior among innovator groups in India. European Journal of Marketing, 38(5/6), 608 – 625.

121. Choumenkovitch, S. F., McKeown, N. M., Tovar, A., Hyatt, R. R., Kraak, V. I., Hastings, A. V., et al. (2013). Whole grain consumption is inversely associated with BMI Z-score in rural school-aged children. Public Health and Nutrition, 16(2), 212e218.

122. Christen WG, Manson JE, Glynn RJ, et al. (2007). β-Carotene supplementation and age-related maculopathy in a randomized trial of US physicians. Arch Ophthalmol.125(3):333-339.

123. Chopra, M., S. Galbraith, and I. Darnton-Hill. (2002). "A Global Response to a Global Problem: The Epidemic of Overnutrition." Bulletin of the World Health Organization 80:952–958.

124. Clarke TB, Davis KM, Lysenko ES, Zhou AY, Yu Y, et al. (2010) Recognition of peptidoglycan from the microbiota by Nod1 enhances systemic innate immunity. Nat Med 16: 228-231.

125. Clements ML, Levine MM, Black RE, et al. (1981). Lactobacillus prophylaxis for diarrhea due to enterotoxigenic Escherichia coli. Antimicrob Agents Chemother 20:104–8.

126. Clifton, P. (2012). Effects of a high protein diet on body weight and comorbidities associated with obesity. British Journal of Nutrition, 108(Suppl. (2)), S122eS129.

127. Clinton SK. (1998). Lycopene: chemistry, biology, and implications for human health and disease. Nutr Rev. 56(2 Pt 1):35-51

128. Clydesdale F (1997). A proposal for the establishment of scientific criteria for health claims for functional foods. Nutrition Review 55: 413–422

129. Clydesdale, Fergus. "Functional foods: opportunities and challenges." Food Tech 58.12 (2004): 35-40

130. Commenges D, Scotet V, Renaud S, Jacqmin-Gadda H, Barberger-Gateau P, Dartigues JF (2000). Intake of flavonoids and risk of dementia. European Journal of Epidemiology 16(4):357-63.

131. Clydesdale F (2004) Functional foods: Opportunities and challenges. Food Technology 58(12) 35–40.

132. Commission of the European Communities. Proposal for a regulation of the European Parliament and of the Council on nutrition and health claims made on foods. Brussels, 2003. http://europa.eu.int/eur-lex/en/com/pdf/2003/com2003_0424en01.pdf (Accessed 21.02.17).

133. Committee on Opportunities in the Nutrition and Food Sciences, Food and Nutrition Board, Institute of Medicine (1994) Enhancing the food supply. Thomas, P. R. Earl, R. eds. Opportunities in the Nutrition and Food Sciences: Research Challenges and the Next Generation of Investigators :98-142 National Academy Press Washington, DC.

134. Consensus Document (1999). Scientific concepts of functional foods in Europe consensus document. Br J Nutr 81, S1–S27.

135. Cosgrove J. (2008). Omega 3 Beverages. Nutraceuticals World. [online]. Available from: http://www.nutraceuticalsworld.com/contents/view_online-exclusives/2008-06-01/omega-3-beverages. (Accessed 19.02.17)

136. Cotter, P.A., Craig, S.R. and McLean, E. (2008). Hyperaccumulation of selenium in hybrid striped bass: a functional food for aquaculture? Aquaculture Nutrition, 14(3), 215-222.

137. Coveney J Bunton R (2003) In pursuit of the study of pleasure: Implications for health research and practice. Health: An Interdisciplinary Journal for the Social Study of Health, Medicine, and Illness 7(2) 161–179.

138. Coudray C, Bellanger J, Castiglia-Delavaud C, Remesy C, Vermorel M, Rayssignuier Y. (1997). Effect of soluble or partly soluble dietary fibres supplementation on absorption and balance of calcium,

magnesium, iron and zinc in healthy young men. Eur J Clin Nutr 51:375–80.

139. Crawford R (1984) A cultural account of "health": Control, release, and the social body. In: Issues on the Political Economy of Health Care (McKinley JB Ed.) pp. 265–308. New York: Tavistock Publications.

140. Crawford R (2006) Health as a meaningful social practice. Health: An Interdisciplinary Journal for the Social Study of Health, Medicine, and Illness 10(4) 401–420.

141. Crawford, M. (2010). Fiber as a functional food. Nutraceuticals World. Available from: http://www.nutraceuticalsworld.com/issues/2010-04/view_features/fiber-as-a-functional-food. (Accessed 13.02.17)

142. Cronin, D.A., Powell, R. and Gormley, T.R. (1991). An examination of the n-3 and n-6 polyunsaturated fatty acid status of wild and farmed salmon (Salmo salar). Irish Journal of Food Science and Technology, 15, 53-62

143. Cummings JH, Macfarlane GT, Englyst HN (2001). Prebiotic digestion and fermentation. Am J Clin Nutr 73(suppl):415S–20S.

144. Cummings JH, Pomare EW, Branch WJ, Naylor CP, Macfarlane GT (1987) Short chain fatty acids in human large intestine, portal, hepatic and venous blood. Gut 28: 1221-1227

145. Cunha, C. A., Lira, F. S., Rosa Neto, J. C., Pimentel, G. D., Souza, G. I. H., et al. (2013). Green tea extract supplementation induces the lipolytic pathway, attenuates obesity, and reduces lowgrade inflammation in mice fed a high-fat diet. Mediators of Inflammation, 635470.

146. Cunnane, SC, Ganguli, S., Menard, C., Liede, AC., Hamadeh, MJ., Chen, ZY., Wolever, TMS.. and Jenkins, DJA. (1993). High-linolenic acid flaxseed (Linum usitatissimum): some nutritional properties in humans. British Journal of Nutrition 69: 443-453.

147. Daglia M, Papetti A, Dacarro C & Gazzani G (1998) Isolation of an antibacterial component from roasted coffee. J Pharm Biomed Anal 18, 219–225.

148. Daglia M, Tarsi R, Papetti A, Grisoli P, Dacarro C, Pruzzo C & Gazzani G (2002) Antiadhesive effect of green and roasted coffee

on Streptococcus mutans' adhesive properties on saliva-coated hydroxyapatite beads. J Agric Food Chem 50, 1225–1229

149. Daglia M, Racchi M, Papetti A, Lani C, Govoni S & Gazzani G (2004) In vitro and ex vivo antihydrosyl radical activity of green and roasted coffee. J Agric Food Chem 52, 1700–1704

150. Dai Z, Wang R, Ang LW, Low YL, Yuan JM, Koh WP. (2014). Protective effects of dietary carotenoids on risk of hip fracture in men: the Singapore Chinese Health Study. J Bone Miner Res 29(2):408-417.

151. Dai, F.-J., Hsu, W.-H., Huang, J.-J., & Wu, S.-C. (2013). Effect of pigeon pea (Cajanus cajan L.) on high-fat diet-induced hypercholesterolemia in hamsters. Food Chemistry and Toxicology, 53, 384e391.

152. Dalar, A., & Konczak, I. (2013). Phenolic contents, antioxidant capacities and inhibitory activities against key metabolic syndrome relevant enzymes of herbal teas from Eastern Anatolia. Industrial Crops Products, 44, 383e390.

153. Dallas, C., Gerbi, A., Tenca, G., Juchaux, F., & Bernard, F. X. (2008). Lipolytic effect of a polyphenolic citrus dry extract of red orange, grapefruit, orange (SINETROL) in human body fat adipocytes. Mechanism of action by inhibition of cAMP-phosphodiesterase (PDE). Phytomedicine, 15(10), 783e792.

154. Dang, ZC. (2009) Dose-dependent effects of soy phyto-oestrogen genistein on adipocytes: mechanisms of action. Obesity Reviews 10(3), pp 342–349

155. Danik Martirosyan and Jaishree Singh: A New Definition of Functional Food by FFC: Creating Functional Food Products Using New Definition. In "Introduction to Functional Food Science", 3-rd edition. Edited by Martirosyan DM, Dallas: Food Science Publisher; 2015:10-24

156. Danthiir, V., Burns, N.R., Nettelbeck, T. and Wilson, C. (2011). The older people, omega-3, and cognitive health (EPOCH) trial design and methodology: A randomised, double-blind, controlled trial investigating the effect of long-chain omega-3 fatty acids on cognitive ageing and wellbeing in cognitively healthy older adults. Nutrition Journal, 10, 117-125

157. Das, S.K., Sanyal, K. and Basu, A. (2005). Study of urban community survey in India: growing

158. trend of high prevalence of hypertension in a developing country. International Journal of Medical Sciences, 2(2), 70–78.

159. Das, S., Raychaudhuri, U., Falchi, M., Bertelli, A., Braga, P. C., & Das, D. K. (2011). Cardioprotective properties of raw and cooked eggplant (Solanum melongena L). Food Function, 2(7), 395e399.

160. David LA, Maurice CF, Carmody RN, Gootenberg DB, Button JE, et al. (2014) Diet rapidly and reproducibly alters the human gut microbiome. Nature 505: 559-563

161. Dean M, Shepard R, Arvola A, Vassallo M, Winkelmann M, Claupein E, Lahteenmaki L, Raats MM, Saba A. (2007). Consumer perceptions of healthy cereal products and production methods. J Cereal Sci. 47:188-196.

162. Deccan Herald (2012). Tata Strategic pegs wellness foods market at Rs 55K crore. Available at: http://www.deccanherald.com/content/61120/tata-strategicpegs-wellness-foods.html [Accessed 22.02.17]

163. De dios Pozo-Olano J, Warram JH Jr, Gomez RG, Cavazos MG. (1978). Effect of a lactobacilli preparation on traveler's diarrhea. A randomized, double blind clinical trial. Gastroenterology 74:829–30.

164. De Graaf J, De Sauvage Nolting PR, Van Dam M, Belsey EM, Kastelein JJ, Haydn Pritchard P, Stalenhoef AF (2002). Consumption of tall oil-derived phytosterols in a chocolate matrix significantly decreases plasma total and low-density lipoprotein-cholesterol levels. British Journal of Nutrition 88(5):479-88

165. Deis, R. (2003). "What's In a Claim: Labeling Functional Foods." Food Product Design November: 15–23. FDA, Department of Health and Human Services "Consumer Health Initiative for Better Nutrition." Task Force Final Report. http: //vm.cfsan.fda.gov/~dms/nuttftoc.html.

166. De Lemos ML (2001). Effects of Soy Phytoestrogens Genistein and Daidzein on Breast Cancer Growth. Annals of Pharmacotherapy 35(9)

167. de Rijk MC1, Breteler MM, den Breeijen JH, Launer LJ, Grobbee DE, van der Meché FG, Hofman A. (1997). Dietary antioxidants and Parkinson disease. The Rotterdam Study. Archives of Neurology. 54(6):762-5.

168. DeFelice, S.L. (1995). 'The nutraceutical revolution, its impact on food industry research and development', Trends Food Sci Technol 6, 59–61

169. DeFuria, J., Bennett, G., Strissel, K. J., Perfield, J. W., II, Milbury, P. E., Greenberg, A. S., et al. (2009).

170. Dietary blueberry attenuates whole-body insulin resistance in high fat-fed mice by reducing adipocyte death and its inflammatory sequelae. Journal of Nutrition, 139(8), 1510e1516.

171. Demeritt L (2008) Consumer Perceptions of Functional Foods. Natural Products Insider. http://www.naturalproductsinsider.com/articles/2008/04/consumer-perceptions-offunctionalfoods.aspx.

172. Denke MA. (1995). Lack of efficacy of low-dose sitostanol therapy as an adjunct to a cholesterol-lowering diet in men with moderate hypercholesterolemia. Am J Clin Nutr. 61:392–396.

173. Denke MA (19995). Lack of efficacy of low-dose sitostanol therapy as an adjunct to a cholesterol-lowering diet in men with moderate hypercholesterolemia. Am J Clin Nutr. 1995, 61: 392-396.

174. Devaraj S, Jialal I, Vega-López S (2004). Plant sterol-fortified orange juice effectively lowers cholesterol levels in mildly hypercholesterolemic healthy individuals. Arterioscler Thrombosis and Vascular Biology 24(3):e25-8

175. Di Mascio, P., Kaiser, S., and Sies, H. (1989). Lycopene as the most efficient biological carotenoid singlet oxygen quencher. Archives of Biochemistry and Biophysics 274, p. 532-538.

176. Dietitians of Canada. What are Functional Foods and Nutraceuticals (2010) Available from: http://www.dietitians.ca/Nutrition-Resources-A-Z/Fact-Sheet-Pages(HTML)/Miscellaneous/Functional-foods-and-nutraceuticals.aspx. (Accessed 22.02.17).

177. Diplock, A.T., Aggett, P.J., Ashwell, M., Bornet, F., Fern, E.B., & Roberfroid, M.B. (1999). Scientific concepts of functional foods in Europe: Consensus document. British Journal of Nutrition, 81, S1–S27.

178. D'Odorico A, Martines D, Kiechl S, et al. (2000). High plasma levels of alpha- and β-carotene are associated with a lower risk of atherosclerosis: results from the Bruneck study. Atherosclerosis. 2000;153(1):231-239

179. Dogasaki C, Shindo T, Furuhata K & Fukuyama M (2002) Identification of chemical structure of antibacterial components against Legionella pneumophila in a coffee beverage. Yakugaku Zasshi 122, 487–494.

180. Dorant, E., van den Brandt, PA., Goldbohm, RA., Hermus, RJJ., and Sturmans, F. (1993). Garlic and its significance for the prevention of cancer in humans: A critical review. British Journal of Cancer 67: 424-429.

181. Doyon, M. and Labrecque, J. (2008). Functional Foods: A Conceptual Definition. British Food Journal, 110(11), 1133-1149

182. Dreosti, IE., Wargovich, MJ., and Yang, CS. (1997). Inhibition of carcinogenesis by tea: The evidence from experimental studies. Critical Reviews in. Food Science and Nutrition 37, p. 761- 770.

183. Dubey J. and Patel R.P. (2004). Small wonders of the Indian market. Journal of Consumer Behaviour, 4(2), 145–151.

184. Duggan C, Penny M, Hibberd P, et al. (2001). The effects of infant cereal supplemented with ologofructose in Peruvian infants: a randomized, double blind trial. J Pediatr Gastroenterol Nutr 33:396

185. Dulloo, A. G., Seydoux, J., Girardier, L., Chantre, P., & Vandermander, J. (2000). Green tea and thermogenesis: Interactions between catechin-polyphenols, caffeine and sympathetic activity. International Journal of Obesity, 24(2), 252e258.

186. Dunne C, O'Mahony L, Murphy L, Thornton G, Morrissey D, O'Halloran S, Feeney M, Flynn S, Fitzgerald G, Daly C, Kiely B, O'Sullivan GC, Shanahan F, Collins JK (2001). In vitro selection criteria for probiotic bacteria of human origin: correlation with in vivo findings. Am J Clin Nutr 73(2 Suppl):386S-392S.

187. Dwyer JH, Paul-Labrador MJ, Fan J, Shircore AM, Merz CN, Dwyer KM (2004). Progression of carotid intima-media thickness and plasma antioxidants: The Los Angeles Atherosclerosis Study. Arterioscler Thromb Vasc Biol. 24(2):313-9. (PubMed)

188. Edirisinghe, I., Banaszewski, K., Cappozzo, J., Sandhya, K., Ellis, C. L., Tadapaneni, R., et al. (2011). Strawberry anthocyanin and its association with postprandial inflammation and insulin. British Journal of Nutrition, 106(6), 913e922.

189. Eichengreen, B. and Gupta, P. (2010). The Service Sector as India's Road to Economic Growth?. Working paper [249], ICRIER, New Delhi, India, April 20.

190. Eichhorn JC, Lee R, Dunster C, Basu S, Kelly FJ (2004). Alpha- and gamma-tocopherol plasma and urinary biokinetics following alpha-tocopherol supplementation. Ann N Y Acad Sci. 1031:339-340.

191. Eilander, A., Hundscheid, D.C., Osendarp, S.J., Transler, C. and Zock, P.l. (2007). Effects of n-3 long chain polyunsaturated fatty acid supplementation on visual and cognitive development throughout childhood: a review of human studies. Prostaglandins, Leukotriens and Essential Fatty Acids, 76, 190-203

192. Eitenmiller RR (1997). Vitamin E content of fats and oils: Nutritional implications. Food Technol 51:78- 81.

193. Ekanayake, P.M., Park, G.T., Lee, Y.D., KIM, S.J., Jeong, S.C. and Le, J. (2005). Antioxidant potential of eel (Anguilla japonica and Conger myriaster) flesh and skin. Journal of Food Lipids, 12(1), 34- 37

194. Elegbede, JA., Maltzman, TH., Elson, CE., and Gould, MN. (1993). Effects of anticarcinogenic monoterpenes on phase II hepatic metabolizing enzymes. Carcinogenesis 14, p. 1221-1223.

195. Eliassen AH, Hendrickson SJ, Brinton LA, et al. (2012). Circulating carotenoids and risk of breast cancer: pooled analysis of eight prospective studies. J Natl Cancer Inst 104(24):1905-1916

196. Ellwood, K.C., Trumbo, P.R. and Kavanaugh, C.J (2009). How the US Food and Drug Administration evaluates the scientific evidence for health claims. Nutrition Reviews 68: 114-121

197. Elson CE, Qureshi AA. (1995). Coupling of cholesterol and tumor-suppressive actions of palm oil to the impact of its minor constituents on 3-hydroxy-3-methylglutaryl coenzyme A reductase activity. Prostaglandins Leukot Essent Fatty Acids 52:205-8.

198. Engelhart MJ, Geerlings MI, Ruitenberg A, van Swieten JC, Hofman A, Witteman JC, Breteler MM. (2002). Dietary intake of

antioxidants and risk of Alzheimer disease. Journal of the American Medical Association 26;287(24):3223-9

199. English, D.R., MacInnis, R.J., Hodge, A.M., Hopper, J.L., Haydon, A.M. and Giles, J.G. (2004). Red meat, chicken and fish consumption and risk of colorectal cancer. Cancer Epidemiology Biomarkers and Prevention, 13(9), 1509-1514
200. Erdman, JW., and Potter, SM. (1997). Soy and bone health. The Soy Connection 5(2), p 1-4.
201. Erlwanger, K. H., Donaldson, J., & Davidson, B. C. (2007). The effects of some over-the-counter dietary weight loss supplements on growth, hepatic glycogen stores, liver lipid profile, pancreatic protein and the gastrointestinal tract of rats. Journal of Biological Science, 7(7), 1201e1207.
202. Ernst, E. (1997). Can allium vegetables prevent cancer? Phytomed. 4, p. 79-83.
203. Etminan M, Takkouche B, Caamano-Isorna F (2004). The role of tomato products and lycopene in the prevention of prostate cancer: a meta-analysis of observational studies. Cancer Epidemiol Biomarkers Prev. 13(3):340-345.
204. Evans J (2008). Antioxidant supplements to prevent or slow down the progression of AMD: a systematic review and meta-analysis. Eye. 2008;22(6):751-760.
205. Evans JR, Henshaw K (2008). Antioxidant vitamin and mineral supplements for preventing age-related macular degeneration. Cochrane Database Syst Rev. CD000253.
206. Evans RW, Shaten BJ, Day BW, Kuller LH. (1998). Prospective association between lipid soluble antioxidants and coronary heart disease in men. The Multiple Risk Factor Intervention Trial. Am J Epidemiol. 147(2):180-186.
207. Fahey, JW., Zhang, Y., and Talalay, P. (1997). Broccoli sprouts: An exceptionally rich source of inducers of enzymes that protect against chemical carcinogens. Proceedings of the National Academy of Sciences 94, p. 10366-10372.
208. Fairus, A., Ima Nirwana, S., Elvy Suhana, M. R., Tan, M. H., Santhana, R., & Farihah, H. S. (2013). Piper sarmentosum is comparable to glycyrrhizic acid in reducing visceral fat deposition

in adrenalectomised rats given dexamethasone. Clinica TerapeuticA, 164(1), 5e10.

209. FAO/WHO/UNU (2002) Protein and Amino Acid Requirements in Human Nutrition. In: Report of a Joint FAO/WHO/UNU Expert Consultation, World Health Org Tech Report No.935

210. Fardet, A. (2010). New hypotheses for the health-protective mechanisms of whole-grain cereals: what is beyond fibre?. Nutrition research reviews, 23(01), 65-134.

211. FDA Food labeling: health claims; plant sterol/stanol esters and coronary heart disease. Food and Drug Admin Fed Reg. 2000 65:54686–54739. http://www.access.gpo.gov/su_docs/fedreg/a000908c.html

212. FDA Regulation (2013). U.S. Food and Drug Administration. Guidance for Industry: Evidencebased review system for the scientific evaluation of health claims – Final. http://www.fda.gov/Food/GuidanceRegulation/GuidanceDocumentsRegulatoryInformation/LabelingNutrition/ucm073332.htm. (Accessed 22.02.17)

213. Feeney J, Finucane C, Savva GM, et al. (2013). Low macular pigment optical density is associated with lower cognitive performance in a large, population-based sample of older adults. Neurobiol Aging. 34(11):2449-2456. (PubMed)

214. Feldman EB. (1999). Assorted monounsaturated fatty acids promote healthy hearts. American Journal of Clinical Nutrition 70:953–954.

215. Fennessy, M.B., Moneley, D.S., Wang, J.H., Kelly, C.J. and Bouchier-Hayes, D.J. (2003). Taurine and vitamin C modify monocyte and endothelial dysfunction in young smokers. Circulation 107(3), 410-415.

216. Fenwick, GR, Heaney RK, Mullin WJ. (1983). Glucosinolates and their breakdown products in food and food plants. Crit Rev Food Sci Nutrititon 18:123-201.

217. Finley, J.W. (2003). Reduction of cancer risk by consumption of selenium-enriched plants: enrichment of broccoli with selenium increases the anticarcinogenic properties of broccoli. Journal of Medicinal Food, 6, 19-26.

218. Floch MH (2005). Use of diet and probiotic therapy in the irritable bowel syndrome: analysis of the literature. J Clin Gastroenterol 39 (5 Suppl): S243-S246.
219. Flood V, Smith W, Wang JJ, Manzi F, Webb K, Mitchell P. (2002). Dietary antioxidant intake and incidence of early age-related maculopathy: the Blue Mountains Eye Study. Ophthalmology. 109(12):2272-2278.
220. Food and Agricultural Organization of the United Nations. Report on Functional Foods; 2007 Available from: http://www.fao.org/ag/agn/agns/files/Functional_Foods_Report_Nov2007.pdf. (Accessed 22.02.17).
221. Food and Agriculture Organisation/World Health Organisation (1998). Carbohydrates in human nutrition: report of a joint FAO/WHO expert consultation, FAO Food and Nutrition paper 66, Rome.
222. Food and Nutrition Board, Institute of Medicine. β-Carotene and other carotenoids. Dietary reference intakes for vitamin C, vitamin E, selenium, and carotenoids. Washington, D.C.: National Academy Press; 2000:325-400.
223. Food in Canada (2010). Yogurt Containing Plant Sterols Launches in Canada. Available from: http://www.foodincanada.com/food-business/yogurt-containing-plant-sterols-launches-in-canada-8269/. Accessed 20.02.17
224. Foster JA, McVey Neufeld KA (2013) Gut-brain axis: how the microbiome influences anxiety and depression. Trends Neurosci 36: 305-312.
225. Gale CR, Hall NF, Phillips DI, Martyn CN (2003). Lutein and zeaxanthin status and risk of age-related macular degeneration. Invest Ophthalmol Vis Sci. 44(6):2461-2465.
226. Gallicchio L, Boyd K, Matanoski G, et al. (2008). Carotenoids and the risk of developing lung cancer: a systematic review. Am J Clin Nutrition 88(2):372-383.
227. Gann PH, Deaton RJ, Rueter EE, et al. (2015). A phase II randomized trial of lycopene-rich tomato extract among men with high-grade prostatic intraepithelial neoplasia. Nutr Cancer. 2015;67(7):1104-1112

228. Garcia-Lafuente, A., Moro, C., Villares, A., Guillamon, E., Rostagno, M. A., D'Arrigo, M., et al. (2011). Mushrooms as a source of anti-inflammatory agents. American Journal of Community Psychology, 48(2-Jan), 125e141.

229. Gärtner C, Stahl W, Sies H (1997). Lycopene is more bioavailable from tomato paste than from fresh tomatoes. Am J Clin Nutr. 66(1):116-122.

230. Gaziano JM, Manson JE, Branch LG, Colditz GA, Willett WC, Buring JE (1995). A prospective study of consumption of carotenoids in fruits and vegetables and decreased cardiovascular mortality in the elderly. Ann Epidemiol. 5(4):255-260. (PubMed)

231. Geremias, R., Pedrosa, R. C., Locatelli, C., De Favere, V. T., CouryPedrosa, R., & Laranjeira, M. C. M. (2006). Lipid lowering activity of hydrosoluble chitosan and association with Aloe vera L. and Brassica olearaceae L. Phytotherapy Research, 20(4), 288e293.

232. Gerster, H. (1997). The potential role of lycopene for human health. Journal of the American College of. Nutrition 16, p.109-126.

233. Gibson GR, Fuller R. (2000). Aspects of in vitro and in vivo research approaches directed toward identifying probiotics and prebiotics for human use. J Nutr 130(suppl):391S–5S.

234. Gibson GR, Roberfroid MB (19995). Dietary modulation of the human colonic microbiota: introducing the concept of prebiotics. J Nutr 125:1401–12

235. Gillies ME & Birkbeck JA (1983) Tea and coffee as sources of some minerals in the New Zealand diet. Am J Clin Nutr 38, 936–942

236. Gillum, R.F., Mussolino, M.E. and Madans, J.H. (1996). The relationship between fish consumption and stroke incidence. The NHANES I epidemiologic follow-up study (National Health and Nutrition Examination Survey). Archives of Internal Medicine, 156(5), 537-542.

237. Giovannucci E (2002). A review of epidemiologic studies of tomatoes, lycopene, and prostate cancer. Exp Biol Med (Maywood). 227(10):852-859.

238. Giovannucci, E., Ascherio, A., Rimm, EB., Stampfer, MJ., Colditz, GA., and Willett, WC. (1995) Intake of carotenoids and retinol in

relation to risk of prostate cancer. Journal of the National Cancer Institute 87, p. 1767-1776.

239. Gomez, J. M., & Guerrero, M. F. (2009). EvaluaciA~3 n de los efectos cardiovasculares del extracto y compuestos de Solanum tuberosum en rata desmedulada [Assessment of the cardiovascular effects of the extract and compounds from Solanum tuberosum in the Pithed rat]. Vitae, 16(3), 320e326.

240. Gontero P, Marra G, Soria F, et al. (2015). A randomized double-blind placebo controlled phase I-II study on clinical and molecular effects of dietary supplements in men with precancerous prostatic lesions. Chemoprevention or "chemopromotion"? Prostate.75(11):1177-1186.

241. Gooda, S. N., Abdul-Hamid, A., Saari, N., Abas, F., Pak Dek, M. S., & Rahim, M. (2012). Anti-pancreatic lipase and antioxidant activity of selected tropical herbs. International Journal of Food Properties, 15(3), 569e578.

242. Gormley, T.R. (2006). Fish as a functional food. Food Science and Technology, 20(3), 25-28. Gormley, T.R. 2008. Overview of seafood research at Ashtown Food Research Centre. In: Downey, G. (ed.) End-of-Project Report. Teagasc Food Research Ashtown, Dublin, Ireland.

243. Gormley, T.R. 2009. Cooperation ISAFRUIT. Food Science and Technology, 23(3), 26-28. Gormley, T.R. 2010a. Inherent functionality: a useful term for consumer information. In: Gormley, T.R. and Holm, F. (eds) Functional Foods: Some Pointers for Success. UCD Institute of Food and Health, University College Dublin, Dublin, Ireland, http://www.ucd.ie/t4cms/ffnet%20funct%20fds%20final%20epublication%20jan%202011.pdf, 216.

244. Gormley, T.R. (2010). Ethics and sustainability in the seafood chain. Food Science and Technology, 24 (3), 14-17.

245. Gormley, T.R., Neumann, T. and Fagan, J.D. (2007). Taurine content of raw and processed fish fillets/portions. European Food Research and Technology, 225, 837-842.

246. Goto, T., Teraminami, A., Lee, J.-Y., Ohyama, K., Funakoshi, K., Kim, Y.-I., et al. (2012). Tiliroside, a glycosidic flavonoid, ameliorates obesity-induced metabolic disorders via activation of adiponectin

signaling followed by enhancement of fatty acid oxidation in liver and skeletal muscle in obese-diabetic mice. Journal of Nutritional Biochemistry, 23(7), 768e776.

247. Gould, MN. (1997). Cancer chemoprevention and therapy by monoterpenes. Environmental Health Perspective 105, p. 977- 979.

248. Gracey, M. (2002). Child health in an urbanizing world. Acta Paediatricia, 91, 1- 8.

249. Graham, HN. (1992). Green tea composition, consumption and polyphenol chemistry. Preventive Medicine 21, p. 334- 350.

250. Greenberg ER, Baron JA, Karagas MR, et al. (1996). Mortality associated with low plasma concentration of β-carotene and the effect of oral supplementation. JAMA. 275(9):699-703.

251. Gremaud G, Dalan E, Piguet C, Baumgartner M, Ballabeni P, Decarli B, Leser ME, Berger A, Fay LB (2002). Effects of non-esterified stanols in a liquid emulsion on cholesterol absorption and synthesis in hypercholesterolemic men. Eur J Nutr. 41(2):54-60

252. Gremaud G, Dalan E, Piguet C, Baumgartner M, Ballabeni P, Decarli B, Leser ME, Berger A, Fay LB (2002). Effects of non-esterified stanols in a liquid emulsion on cholesterol absorption and synthesis in hypercholesterolemic men. European Journal of Nutrition 41(2):54-60

253. Gremaud G, Dalan E, Piguet C, Baumgartner M, Ballabeni P, Decarli B, Leser ME, Berger A, Fay LB (2002). Effects of non-esterified stanols in a liquid emulsion on cholesterol absorption and synthesis in hypercholesterolemic men. Eur J Nutr. 41: 54-60.

254. Gremaud G, Piguet C, Baumgartner M, Pouteau E, Decarli B, Berger A, Fay LB (2001). Simultaneous assessment of cholesterol absorption and synthesis in humans using on-line gas chromatography/ combustion and gas chromatography/pyrolysis/isotope-ratio mass spectrometry. Rapid Commun Mass Spectrom. 15(14):1207-13.

255. Gremaud G, Piguet C, Baumgartner M, Pouteau E, Decarli B, Berger A, Fay LB(2001). Simultaneous assessment of cholesterol absorption and synthesis in humans using on-line gas chromatography/ combustion and gas chromatography/pyrolysis/isotope-ratio mass spectrometry. Rapid Commun Mass Spectrom 15(14):1207-13

256. Grice, E.A. & Segre, J.A. (2011). The skin microbiome. Nature Reviews Microbiology, 9, 244-253. doi: 10.1038/nrmicro2537 The Human Microbiome Project Consortium (2012). Structure, function and diversity of the healthy human microbiome. Nature, 486, 207-214. doi: 10.1038/nature11234

257. Grice, E.A., Kong, H.H., Conlan, S., Deming, C.B., Davis, J., Young, A.C., NISC Comparative Sequencing Program, Bouffard, G.G., Blakesley, R.W., Murray, P.R., Green, E.D., Turner, M.L. & Segre, J.A. (2009). Topographical and temporal diversity of the human skin microbiome. Science, 324(5931), 1190-1192. doi: 10.1126/science.1171700

258. Griffiths, P.L. and Bentley M.E. (2001). The Nutrition Transition Is Underway in India. The Journal of Nutrition, 131(10), 2692-2700

259. Grundy SM & Mok HY (1976) Effects of low dose phytosterols on cholesterol absorption in man. In Lipoprotein Metabolism, pp. 112–118 [H Greten, editor]. Berlin, Heidelberg and New York: Springer

260. Gu, Y., Yu, S., & Lambert, J. D. (2013). Dietary cocoa ameliorates obesity-related inflammation in high fat-fed mice. European Journal of Nutrition, 1e10. Hanwell, H. E. C., Kay, C. D., Lampe, J. W., Holub, B. J., & Duncan, A. M. (2009).

261. Guarino A, Lo Vecchio A, Canani RB (2009) Probiotics as prevention and treatment for diarrhea. Curr Opin Gastroenterol 25: 18-23

262. Gylling H, Puska P, Vartiainen P & Miettinen TA (1996) Serum retinol a-tocopherol, carotenes and lipid peroxide production during serum cholesterol lowering by sitostanol ester margarine in a mildly hypercholesterolemic population. Circulation 94, Suppl. 1, 578.

263. Hak AE, Stampfer MJ, Campos H, et al. (2003). Plasma carotenoids and tocopherols and risk of myocardial infarction in a low-risk population of US male physicians. Circulation. 108(7):802-807. (PubMed)

264. Hallikainen MA, Sarkkinen ES, Uusitupa MI (2000). Plant stanol esters affect serum cholesterol concentrations of hypercholesterolemic men and women in a dose-dependent manner. Journal of Nutrition 130(4):767-776

265. Hammond BR, Jr., Johnson EJ, Russell RM, et al. (1997). Dietary modification of human macular pigment density. Invest Ophthalmol

Vis Sci. 38(9):1795-1801.

266. Harbowy, ME. and Balentine, DA. (1997). Tea Chemistry. Critical Reviews in Plant Science 16, p. 415-480.

267. Hardy, G. (2000). Nutraceuticals and functional foods: Introduction and meaning. Nutrition, 16, 688–697.

268. Hartstra AV, Bouter KE, Bäckhed F, Nieuwdorp M (2015) Insights into the role of the microbiome in obesity and type 2 diabetes. Diabetes Care 38: 159-165.

269. Hasegawa, S. and Miyake, M. (1996). Biochemistry and biological functions of citrus limonoids. Food Reviews International. 12, p. 413-435.

270. Hasler C (2000) 3e changing face of functional foods. Journal of the American College of Nutrition 19(5) 499s–506s.

271. Hasler C Brown A (2009) Position of the American Dietetic Association: Functional foods. Journal of the American Dietetic Association 109(4) 735–746.

272. Hasler CM, Brown AC 92009). Position of the American Dietetic Association: Functional foods. Journal of the American Dietetic Association 09: 735-746.

273. Hasler CM (2002). Functional Foods: Benefits, Concerns and Challenges—A Position Paper from the American Council on Science and Health. The Journal of Nutrition 132: 3772-3781.

274. Hasler, CM. (1998). Functional Foods: Their Role in Disease Prevention and Health Promotion. Food Technology 52(2), p. 57-62.

275. Hayes, K. C., Pronczuk, A., & Perlman, D. (2004). Nonesterified phytosterols dissolved and recrystallized in oil reduce plasma cholesterol in gerbils and humans. The Journal of nutrition, 134(6), 1395-1399.

276. Health Canada. Policy paper – Nutraceuticals/functional foods and health claims on foods; 1998 Available from: http://www.hc-sc.gc.ca/fn-an/labeletiquet/claims-reclam/nutra-funct_foods-nutra-fonct_aliment-eng.php. (Accessed 22.02.17).

277. Health Canada. Policy Options Analysis: Nutraceuticals/Functional Foods, Health Canada, Health Protection Branch, Therapeutic

Products Programme and Food Directorate, Ottawa, 1997

278. Health Council of the Netherlands. Foods and dietary supplements with health claims. The Hague: Health Council of the Netherlands, 2003. (Publication no. 2003/09E.)

279. Heaney RK, Fenwick GR. Natural toxins and protective factors in Brassica species, including rapeseed. Nat Toxins. 1995; 3:233-7.

280. Heasman M: The regulation of functional foods and beverages in Japan. In Proceedings of the first Vitafoods International Conference. Edited by Blenford DE. Copenhagen: Food Tech Europe; 1997.

281. Hecht, SS. (1995). Chemoprevention by isothiocyanates. Journal of Cellular Biochemistry. Suppl. 22, p. 195-209.

282. Heinemann T, Axtmann G & von Bergmann K (1993) Comparison of intestinal absorption of

283. cholesterol with different plant sterols in man. European Journal of Clinical Investigation 23, 827–831

284. Heinemann T, Leiss O, von Bergmann K (1986). Effect of low-dose sitostanol on serum cholesterol in patients with hypercholesterolemia. Atherosclerosis. 1986, 61: 219-223.

285. Hendler SS, Rorvik DM, eds. PDR for Nutritional Supplements. 2nd ed. Thomson Reuters; 2008.

286. Hendriks HF, Weststrate JA, van Vliet T, Meijer GW (1999). Spreads enriched with three different levels of vegetable oil sterols and the degree of cholesterol lowering in normocholesterolaemic and mildly hypercholesterolaemic subjects. Eur J Clin Nutr. 53:319–327.

287. Hendriks, H. F. J., Weststrate, J. A., Van Vliet, T., & Meijer, G. W. (1999). Spreads enriched with three different levels of vegetable oil sterols and the degree of cholesterol lowering in normocholesterolaemic and mildly hypercholesterolaemic subjects. European Journal of Clinical Nutrition, 53(4), 319-327

288. Hennekens CH, Buring JE, Manson JE, et al. (1996). Lack of effect of long-term supplementation with β-carotene on the incidence of malignant neoplasms and cardiovascular disease. N Engl J Med. 334(18):1145-1149. (PubMed)

289. Hertog, MGL., Feskens, EJM, Hollman, PCH, Katan, MB., and Krumhout, D. (1993). Dietary antioxidant flavonoids and risk of

coronary heart disease: The Zutphen Elderly Study. The Lancet 342, p. 1007-1011.

290. Hirano, S., Itakura, C., Seino, H., Akiyama, Y., Nonaka, I., Kanbara, N. and Kawakami, T. (1990). Chitosan as an ingredient for domestic feeds. Journal of Agriculture and Food Chemistry, 38, 1214-1217.

291. Hodgson, JM., Puddey, IB., Beilin, LJ., Mori, TA., and Croft, KD. (1998). Supplementation with isoflavonoid phytoestrogens does not alter serum lipid concentrations: A randomized controlled trial in humans. Journal of Nutrition 128, p. 728-732.

292. Hoefel, A. L., Hansen, F., Rosa, P. D., De Assis, A. M., Da Silveira, S. L., Denardin, C. C., et al. (2011). The effects of hypercaloric diets on glucose homeostasis in the rat: Influence of saturated and monounsaturated dietary lipids. Cell Biochemistry and Function, 29(7), 569e576.

293. Homma Y, Ikeda I, Ishikawa T, Tateno M, Sugano M, Nakamura H (2003). Decrease in plasma low-density lipoprotein cholesterol, apolipoprotein B, cholesteryl ester transfer protein, and oxidized low-density lipoprotein by plant stanol ester-containing spread: a randomized, placebo-controlled trial. Nutrition 19(4):369-74

294. Hooper L, Kroon PA, Rimm EB, Cohn JS, Harvey I, et al. (2008) Flavonoids, flavonoid-rich foods, and cardiovascular risk: a meta-analysis of randomized controlled trials. Am J Clin Nutr 88: 38-50.

295. Hossain, A., Ishihara, T., Hara, K., Osatomi, K., Khan, A. and Nozahi, Y. (2003). Effect of proteolitic squid protein hydrolysate on the state of water and dehydration-induced denaturation of lizard fish myofibrillar protein. Journal of Agricultural and Food Chemistry, 51(16), 4769-4774.

296. Hosoya N (1998). Health claims in Japan—foods for specified health uses and functional foods. J Nutr Food 1, 1–11.

297. Hoveyda N, Heneghan C, Mahatani KR (2009). A systematic review and meta-analysis: probiotics in the treatment of irritable bowel syndrome. BMC Gastroenterol. 9: 15-10

298. Howell TJ, MacDougall DE, Jones PJ (1998). Phytosterols partially explain differences in cholesterol metabolism caused by corn or olive oil feeding. J Lipid Res. 39: 892-900

299. Hu FB (2003). Plant-based foods and prevention of cardiovascular disease: an overview. Am J Clin Nutr 78(suppl): S544-51.

300. Huang M-T, Ferraro T, Ho C-T. Cancer chemoprevention by phytochemicals in fruits and vegetables:

301. An overview. In: Huang M-T, Osawa T, Ho C-T, Rosen RT (editors). Food phytochemicals for cancer prevention I. Fruits and vegetables. Washington DC: American Chemical Society; 1994. p. 2-16.

302. Huang, T., Sinclair, A.J., Shen, L.R., Yang, B. and Li, D. (2009). Comparative effects of tuna oil and salmon oil on liver lipid 27 metabolism and fatty acid concentration in rats. Journal of Food Lipids, 16(4), 436-451.

303. Hughes CL (1988). Phytochemical mimicry of reproductive hormones and modulation of herbivore fertility by phytoestrogens. Environmental Health Perspectives. 78: 171–174.

304. Hughes N (2009). Health claims thrown into disarray by EFSA rejections. The Grocer, 10 October: 12–13.

305. Huppert M, Macpherson DA, Cazin J (1953) Pathogenesis of Candida albicans infection following antibiotic therapy. I. The effect of antibiotics on the growth of Candida albicans. Journal of Bacteriology 65: 171-176.

306. Husain, M.M. and Nemeroff, C.B. (1990). Neuro-degeneration and disturbance in memory cognition. Journal of the American Geriatric Society, 38 (8), 918-925.

307. Hwang J, Sevanian A, Hodis HN, Ursini F. (2000). Synergistic inhibition of LDL oxidation by phytoestrogens and ascorbic acid. Free Radic Biol Med. 29(1):79-89

308. Ikeda I, Tanabe Y & Sugano M (1989) Effects of sitosterol and sitostanol on micellar solubility of cholesterol. Journal of Nutritional Science and Vitaminology 35, 361–369

309. Impellitteri CA, Allen HE, Lagos G & McLaughlin MJ (2000). Removal of soluble Cu and Pb by the automatic drip coffee brewing process: application to risk assessment. Hum Ecolog Risk Assess 6, 313–322.

310. Ingram, D., Sanders, K., Kolybaba, M., Lopez, D. (1997). Case-control study of phyto-oestrogens and breast cancer. The Lancet. 350 (9083), p990–994

311. International Food Information Council. (2007). Consumer Attitudes toward Functional Foods/Foods for Health. Available from: http://www.foodinsight.org/Content/6/IFICExecSumSINGLE_vF2.pdf. (Accessed 22.02.17).

312. International Life Sciences Institute (1999) Safety assessment and potential health benefits of food components based on selected scientific criteria. ILSI North America Technical Committee on Food Components for Health Promotion. Crit. Rev. Food Sci. Nutr. 39:203-316.

313. Ip, C., Birringer, M., Kotrebal, M., Tyson, J.F., Uden, P.C. and Lisk, D.J. (2000). Chemical speciation influences comparative activity of selenium-enriched garlic and yeast in mammary cancer prevention. Journal of Agriculture and Food Chemistry, 48, 2062-2070.

314. Iribarren C, Folsom AR, Jacobs DR, Jr., Gross MD, Belcher JD, Eckfeldt JH (1997). Association of serum vitamin levels, LDL susceptibility to oxidation, and autoantibodies against MDA-LDL with carotid atherosclerosis. A case-control study. The ARIC Study Investigators. Atherosclerosis Risk in Communities. Arterioscler Thromb Vasc Biol. 17(6):1171-1177.

315. Isaacsohn, JL., Moser, M., Stein, EA., Dudley, K., Davey, JA., Liskov, E., and Black, HR. (1998). Garlic powder and plasma lipids and lipoproteins: A multicenter, randomized, placebo-controlled trial. Archives of Internal Medicine 158, p.1189¬1194.

316. István, S., Emese, K., Beáta, K. and Andrea, L. (2008). Functional food. Product development, marketing and consumer acceptance—A review, Appetite, 51, 456–467

317. Ito M, Kimura M, Deguchi Y, Miyamori-Watabe A, Yajima T, Kan T. (1993). Effects of transgalactosylated disaccharides on the human intestinal microflora and their metabolism. J Nutr Sci Vitaminol (Tokyo) 39:279–88.

318. Ito Y, Kurata M, Suzuki K, Hamajima N, Hishida H, Aoki K. (2006). Cardiovascular disease mortality and serum carotenoid levels: a Japanese population-based follow-up study. J Epidemiol. 16(4):154-160.

319. Jackson KG, Taylor GR, Clohessy AM, Williams CM. (1999). The effect of the daily intake of inulin on fasting lipid, insulin and

glucose concentrations in middle-aged men and women. Br J Nutr 82:23–30.

320. Jackson, P.A., Reay, J.L., Scholey, A.B. and Kennedy. D.O. (2012). DHA-rich fish oil modulates the cerebral hemodynamic response to cognitive tasks in healthy young adults. Biological Psychology, 89, 183-190.

321. Jacobs, E.T. and 14 co-authors. (2004). Selenium and colorectal adenoma: results of a pooled analysis. Journal of the National Cancer Institute, 96, 1669-1675.

322. Jakobsson, H.E., Jernberg, C., Andersson, A.F., Sjolund-Karlsson, M., Jansson, J.K. & Engstrand, L. (2010). Short-term antibiotic treatment has differing long-term impacts on the human throat and gut microbiome. PLoS ONE, 5(3), e9836. doi: 10.1371/journal.pone.0009836

323. Jalal F, Nesheim MC, Agus Z, Sanjur D, Habicht JP (1998). Serum retinol concentrations in children are affected by food sources of β-carotene, fat intake, and anthelmintic drug treatment. Am J Clin Nutr. 68(3):623-629.

324. Jalil, A. M. M., Ismail, A., Pei, C. P., Hamid, M., & Kamaruddin, S. H. S. (2008). Effects of cocoa extract on glucometabolism, oxidative stress, and antioxidant enzymes in obese-diabetic (Ob-db) rats. Journal of Agricultural and Food Chemistry, 56(17), 7877e7884.

325. Jaroonvesama, N., & Attatippaholkun, M. (1991). Treatment of hypertension with Chinese celery. Siriraj Medical Journal, 43(2), 85e89.

326. Jayamurthy, P., Aparna, B., Gayathri, G., & Nisha, P. (2013). Evaluation of antioxidant potential of inflorescence and stalk of plantain (Musa Sapientum). Journal of Food Biochemistry, 37(1), 2e7.

327. Je, J.Y., Qian, Z.J., Lee, S.H., Byun, H.G., and Kim, S.K. (2008). Purification and antioxidant properties of Bigeye tuna (Thunnus obesus) dark muscle peptide on free radical-mediated oxidative systems. Journal of Medicinal Food, 11(4), 629-637.

328. Jeffery E (2005). Component interactions for ef- ficacy of functional foods. J Nutr. 2005;135(5):1223-1225

329. Jenkins DJ, Kendall CW, Marchie A, Faulkner D, Vidgen E, Lapsley

KG, Trautwein EA, Parker TL, Josse RG, Leiter LA, Connelly PW. (2003). The effect of combining plant sterols, soy protein, viscous fibers, and almonds in treating hypercholesterolemia. Metabolism. 2003;52:1478–1483.

330. Jenkins DJ, Kendall CW, Marchie A, Jenkins AL, Connelly PW, Jones PJ, Vuksan V (2003). The Garden of Eden--plant based diets, the genetic drive to conserve cholesterol and its implications for heart disease in the 21st century. Comp Biochem Physiol A Mol Integr Physiol. 136:141–151.

331. Jeon, S.-M., Kim, J.-E., Shin, S.-K., Kwon, E.-Y., Jung, U. J., Baek, N.-I., et al. (2013). Randomized double-blind placebo-controlled trial of powdered brassica rapa ethanol extract on alteration of body composition and plasma lipid and adipocytokine profiles in overweight subjects. Journal of Medicinal Food, 16(2), 133e138.

332. Jodral-Segado AM, Navarro-Alarcon M, Lopez-Ga de la Serrana H & Lopez-Martinez MC (2003)

333. Magnesium and calcium contents in foods from SE Spain: influencing factors and estimation of daily dietary intakes. Sci Total Environ 312, 47–58

334. Johnson EJ, McDonald K, Caldarella SM, Chung HY, Troen AM, Snodderly DM (2008). Cognitive findings of an exploratory trial of docosahexaenoic acid and lutein supplementation in older women. Nutr Neurosci. 11(2):75-83.

335. Johnson EJ, Vishwanathan R, Johnson MA, et al. (2013). Relationship between Serum and Brain Carotenoids, alpha-Tocopherol, and Retinol Concentrations and Cognitive Performance in the Oldest Old from the Georgia Centenarian Study. J Aging Res. 2013:951786.

336. Jones PJ, Howell T, MacDougall DE, Feng JY, Parsons W (1998). Short-term administration of tall oil phytosterols improves plasma lipid profiles in subjects with different cholesterol levels. Metabolism. 47: 751-756.

337. Jones PJ, Ntanios FY, Raeini-Sarjaz M, Vanstone CA (1999). Cholesterol-lowering efficacy of a sitostanol-containing phytosterol mixture with a prudent diet in hyperlipidemic men. Am J Clin Nutr. 69: 1144-1150.

338. Jones PJ, Vanstone CA, Raeini-Sarjaz M, St-Onge MP (2003).

Phytosterols in low- and non-fat beverages as part of a controlled diet fail to lower plasma lipid levels. Journal of Lipid Research 44(9):1713-9

339. Jumpertz R, Le DS, Turnbaugh PJ, Trinidad C, Bogardus C, et al. (2011). Energy-balance studies reveal associations between gut microbes, caloric load, and nutrient absorption in humans. American Journal of Clinical Nutrition 94: 58-65

340. Jungbauer, A., & Medjakovic, S. (2012). Anti-inflammatory properties of culinary herbs and spices that ameliorate the effects of metabolic syndrome. Maturitas, 71(3), 227e239.

341. Kalmijn, S., van Boxtel, M.P.J., Ocke, M., Verschuren, D., Kromhout, D. And Launer, L.J. 2004. Dietary intake of fatty acids and fish in relation to cognitive performance at middle age. Neurology, 2(2), 275-280.

342. Kalra E.K. Nutraceutical -Definition and Introduction. AAPS PharmSci. 2003 5:25. Available online: http://www.aapsj.org/view.asp?art=ps05032(Accessed on 22 February 2017)

343. Kalra, Ekta K. "Nutraceutical-definition and introduction." Aaps Pharmsci 5.3 (2003): 27-28

344. Kamangar F, Qiao YL, Yu B, et al. Lung cancer chemoprevention: a randomized, double-blind trial in Linxian, China. Cancer Epidemiol Biomarkers Prev. 2006;15(8):1562-1564. (PubMed)

345. Kamisoyama, H., Honda, K., Tominaga, Y., Yokota, S., & Hasegawa, S. (2008). Investigation of the anti-obesity action of licorice flavonoid oil in diet-induced obese rats. Bioscience, Biotechnology and Biochemistry, 72(12), 3225e3231.

346. Kanamoto, Y., Yamashita, Y., Nanba, F., Yoshida, T., Tsuda, T., Fukuda, I., et al. (2011). black soybean seed coat extract prevents obesity and glucose intolerance by up-regulating uncoupling proteins and down-regulating inflammatory cytokines in high-fat diet-fed mice. Journal of Agricultural and Food Chemistry., 59(16), 8985e8993.

347. Kang DW, Park JG, Ilhan ZE, Wallstrom G, Labaer J, et al. (2013) Reduced incidence of Prevotella and other fermenters in intestinal microflora of autistic children. PLoS One 8: e68322.

348. Kang JH, Ascherio A, Grodstein F. Fruit and vegetable consumption

and cognitive decline in aging women. Ann Neurol. 2005;57(5):713-720. (PubMed)

349. Kaput, J. and R. Rodriguez. 2004. "Nutritional Genomics: The Next Frontier in the Postgenomic Era." Physiology Genomics 16:166–177. Le Coutre, J. 2003. "Taste: The Metabolic Sense."

350. Katiyar, SK. and Mukhtar, H. (1996). Tea in chemoprevention of cancer: Epidemiologic and experimental studies (review). International Journal of Oncology 8, p. 221-238.

351. Keilen, V., Fagan, J.D. and Gormley, T.R. 2007. Apple pomace as an antioxidant dietary fibre in mackerel products. Poster Presenation. World Seafood Congress, Dublin, Ireland.

352. Keilen, V., Fagan, J.D. and Gormley, T.R. 2007. Apple pomace as an antioxidant dietary fibre in mackerel products. Poster Presenation. World Seafood Congress, Dublin, Ireland.

353. Kellogg, J., Wang, J., Flint, C., Ribnicky, D., Kuhn, P., De Mejia, E. G., et al. (2010). Alaskan wild berry resources and human health under the cloud of climate change. Journal of Agricultural and Food Chemistry, 58(7), 3884e3900.

354. Key TJ, Appleby PN, Travis RC, et al. Carotenoids, retinol, tocopherols, and prostate cancer risk: pooled analysis of 15 studies. Am J Clin Nutr. 2015;102(5):1142-1157. (PubMed)

355. Khan, A.A., Hossain, A., Hara, K., Osatomi, K., Ishihara, T., Osako, K. and Nozaki, Y. 2004. Concentration dependent effect of enzymatic fish protein hydrolysates on the state of water and denaturation of lizard fish (Saurida wanieso) myofibrils during dehydration. Food Science and Technology Research 10(2), 132- 136.

356. Khan, V., Najmi, A. K. M. M. M., & Pillai, K. K. (2012). A pharmacological appraisal of medicinal plants with antidiabetic potential. Journal of Pharmacy and Bioallied Sciences, 4, 27e42.

357. Kianbakht, S., & Hajiaghaee, R. (2011). Anti-hyperglycemic effects of saffron and its active constituents, crocin and safranal, in alloxaninduced diabetic rats. Journal of Medicinal Plant, 10(39), 82e89.

358. Kim, K., Kim, H., Kwon, J., Lee, S., Kong, H., Im, S.-A., et al. (2009). Hypoglycemic and hypolipidemic effects of processed Aloe vera gel in a mouse model of non-insulin-dependent diabetes

mellitus. Phytomedicine, 16(9), 856e863.

359. Kim, K.-J., & Lee, B.-Y. (2012). Fucoidan from the sporophyll of Undaria pinnatifida suppresses adipocyte differentiation by inhibition of inflammation-related cytokines in 3T3-L1 cells. Nutrition Research, 32(6), 439e447.

360. Kim, S.-h., Park, H.-S., Lee, M.-s., Cho, Y.-J., Kim, Y.-S., Hwang, J.-T., et al. (2008). Vitisin a inhibits adipocyte differentiation through cell cycle arrest in 3T3-L1 cells. Biochemical and Biophysical Research Communication, 372(1), 108e113.

361. Kinman, G. and Jones, F. (2005). Lay Representations of Workplace Stress: What Do People Really Mean When They Say They Are Stressed?. Work & Stress, 19(2), 101-120.

362. Kinsella, J.E. 1986. Food components with potential therapeutic benefits: The n-3 polyunsaturated fatty acids of fish oils. Food Technology 28, 89-97.

363. Kitts, D.D. and Weiler, K. 2003. Bioactive proteins and peptides from food sources: Applications of bioprocesses used in isolation and recovery. Current Pharmaceutical Design, 9(16), 1309-1323.

364. Knekt, P., Kumpulainen, J., Järvinen, R., Rissanen, H., Heliövaara, M., Reunanen, A., Hakulinen, T. and Aromaa, A. (2002) Flavonoid intake and risk of chronic diseases. American Journal of Clinical Nutrition 76(3):560-8.

365. Koenig, J.E., Spor, A., Scalfone, N., Fricker, A.D., Stombaugh, J., Knight, R., Angenent, L.T. & Ley, R.E. (2011). Succession of microbial consortia in the developing infant gut microbiome. Proceedings of the National Academy of Sciences of the United States of America, 108(S1), 4578-4585. doi: 10.1073/pnas.1000081107

366. Kohlmeier, L., Kark, JD., Gomez-Gracia, E., Martin, BC., Steck, SE., Kardinaal, AFM., Ringstad, J., Thamm, M., Masaev, V., Riemersma, R., Martin-Moreno, JM., Huttunen, JK., and Kok, FJ. (1997).

367. Lycopene and myocardial infarction risk in the EURAMIC study. American Journal of Epidemiology 146, p. 618-626.

368. Kohlmeier, L., Weerings, KGC, Steck, S., and Kok, FJ. (1997b). Tea and cancer prevention -¬ An evaluation of the epidemiologic literature. Nutrition and Cancer 27, p. 1-13.

369. Kolanowski, W. and Laufenberg, G. 2006. Enrichment of food products with polyunsaturated fatty acids by fish oil addition. European Journal of Food Research and Technology, 22(3-4), 472-477.

370. Kolanowski, W., Swiderski, F, and Berger, S. 1999. Possibilities of fish oil application for food product enrichment with omega-3 PUFA. International Journal of Food Science and Nutrition, 50 (1), 39-49.

371. Kolanowski, W., Swiderski, F., Jaworska, D. and Berger, S. 2004. Stability, sensory quality, texture properties and nutritional value of fish oil-enriched spreadable fat. Journal of the Science of Food and Agriculture 84(15), 2135-2141.

372. Komiyama K, Hayashi M, Cha S, Yamaoka M. Antitumor and antioxidant activity of tocotrienols. In: Ong ASH, Packer L (editors). Lipid-soluble antioxidants: Biochemistry and clinical applications. Basel, Switzerland: Birkhauser Verlag; 1992. p. 152- 9.

373. Kostic AD, Xavier RJ, Gevers D (2014) He microbiome in inflammatory bowel disease: current status and the future ahead. Gastroenterology146: 1489-1499.

374. Kotilainen, L., Rajalahti, R., Ragasa, C. and Pehu, E. (2006). Health enhancing foods: Opportunities for strengthening the sector in developing countries. Agriculture and Rural Development Discussion Paper 30, World Bank, Washington, DC.

375. Krammer HJ, Schlieger F, Harder H: Probiotics as therapeutic agents in irritable bowel syndrome. Z Gastroenterol. 2005, 43 (5): 467-471.

376. Krikorian, R., Nash, T. A., Shidler, M. D., Shukitt-Hale, B., & Joseph, J. A. (2010). Concord grape juice supplementation improves memory function in older adults with mild cognitive impairment. British Journal of Nutrition, 103(5), 730e734.

377. Kris-Etherton PM, Keen CL (2002) Evidence that the antioxidant flavonoids in tea and cocoa are beneficial for cardiovascular health. Curr Opin Lipidol 13: 41-49.

378. Krishnaswamy, K. (1996). Indian functional foods: role in prevention of cancer. Nutrition Reviews, 54(11), S127

379. Kritchevsky SB. β-Carotene, carotenoids and the prevention of coronary heart disease. J Nutr. 1999;129(1):5-8. (PubMed)

380. Krumbein, A., Schonhof, I., & Schreiner, M. (2005). Composition and contents of phytochem s (glucosinolates, carotenoids and chlorophylls) and ascorbic acid in selected Brassica species (B. juncea, B. rapa subsp. nipposinica var. chinoleifera, B. rapa subsp. chinensis and B. rapa subsp. rapa). Journal of Applied Botany and Food Quality, 79(3), 168e174.

381. Krutulyte R, Grunert K, Scholderer J, Skov Hagemann K, Elgaard P, Nielsen B, Graverholt JP. Motivational factors for consuming omega-3 PUFAs: An exploratory study with Danish consumers. Appetite. 2008;51:137-147

382. Kumar NB, Besterman-Dahan K, Kang L, et al. Results of a randomized clinical trial of the action of several doses of lycopene in localized prostate cancer: administration prior to radical prostatectomy. Clin Med Urol. 2008;1:1-14. (PubMed)

383. Kural, B. V., Kucuk, N., Yucesan, F. B., & Orem, A. (2011). Effects of kale (Brassica oleracea L. var. acephala DC) leaves extracts on the susceptibility of very low and low density lipoproteins to oxidation. Indian Journal of Biochemistry and Biophysics, 48(5), 361e364.

384. Kurzer MS (2003). Phytoestrogen supplement use by women. Journal of Nutrition 133(6):1983S-1986S

385. Kwon, Y.-I., Apostolidis, E., & Shetty, K. (2008). In vitro studies of eggplant (Solanum melongena) phenolics as inhibitors of key enzymes relevant for type 2 diabetes and hypertension. Bioresource Technology, 99(8), 2981e2988.

386. Kwon, Y.-I., Apostolidis, E., Kim, Y.-C., & Shetty, K. (2007). Health benefits of traditional corn, beans, and pumpkin: In vitro studies for hyperglycemia and hypertension management. Journal of Medicinal Food, 10(2), 266e275.

387. Kyrø, C., Zamora-Ros, R., Scalbert, A., Tjønneland, A., Dossus, L., Johansen, C., Bidstrup, PE., Weiderpass, E., Christensen, J., Ward, H., Aune, D., Riboli, E., His, M., Clavel-Chapelon F5,6,13, Baglietto L5,6,13,14,15, Katzke V16, Kühn T16, Boeing H17, Floegel A17, Overvad K18, Lasheras C19, Travier N20, Sánchez MJ21,22, Amiano P22,23, Chirlaque MD22,24,25, Ardanaz E22,26,27, Khaw

KT28, Wareham N29, Perez-Cornago A30, Trichopoulou A31,32, Lagiou P31,32, Vasilopoulou E31,32,

388. Masala G33, Grioni S34, Berrino F34, Tumino R35, Sacerdote C36, Mattiello A37, Bueno-de-Mesquita HB12,38,39,40, Peeters PH12,41, van Gils C41, Borgquist S42, Butt S43, Zeleniuch-Jacquotte A44, Sund M45, Hjartåker A46, Skeie G8, Olsen A4, Romieu I3. Pre-diagnostic polyphenol intake and breast cancer survival: the European Prospective Investigation into Cancer and Nutrition (EPIC) cohort. Breast Cancer Res Treat. 2015 Nov;154(2):389-401. doi: 10.1007/s10549-015-3595-9. Epub 2015 Nov 3.
389. Lajolo, M.F. (2002). Functional foods: Latin American perspectives. British Journal of Nutrition, 88(2), S145–S150
390. Lanzotti, V. (2006). The analysis of onion and garlic. Journal of Chromatography A, 1112(2-Jan), 3e22.
391. Laparra J.M., Sanz Y. Interactions of gut microbiota with functional food components and nutraceuticals. Pharmacol. Res. 2010;61:219–225
392. Larsen, E.H., Lobinski, R., Burger-Meijer, K., Hansen, M., Ruzik, R., Mazurowska, L., Rasmussen, P.H., Sloth, J.J., Scholten, O.E. and Kik, C. 2006. Uptake and speciation of selenium in garlic cultivated in soil amended with symbiotic fungi (mycorrhiza) and selenate. Analytical and Bioanalytical Chemistry, 385, 1098-1108.
393. Lauden R (2001) A plea for culinary modernism: Why we should love new, fast, processed food. Gastronomica 1(1) 36–44
394. Lawson, LD., Wang, ZYJ., and Hughes, BG. (1991). Identification and HPLC quantitation of the sulfides and dialk(en)ylthiosulfinates in commercial garlic products. Planta Medica 57, p. 363-370.
395. Lecour, S., & Lamont, K. T. (2011). Natural polyphenols and cardioprotection. Mini-Review in Medicinal Chemistry, 11(14), 1191e1199.
396. Lee YK, Mazmanian SK (2010) Has the microbiota played a critical role in the evolution of the adaptive immune system? Science 330: 1768-1773.
397. Lee, H., Park, D., & Yoon, M. (2013). Korean red ginseng (Panax ginseng) prevents obesity by inhibiting angiogenesis in high fat diet-

induced obese C57BL/6J mice. Food Chemistry and Toxicology, 53, 402e408.

398. Lee, Suan Chua and Foo, Mung Hsia. (2014). Functional Foods and Its Biomarkers. Introduction to Functional Food Science: Textbook. 2nd ed. Richardson, TX: Functional Food Center

399. Lee, W., Eom, D.-W., Jung, Y., Yamabe, N., Lee, S., Jeon, Y., et al. (2012). Dendrobium moniliforme attenuates high-fat diet-induced renal damage in mice through the regulation of lipid-induced oxidative stress. American Journal of Chinese Medicine, 40(6), 1217e1228.

400. Leenders M, Leufkens AM, Siersema PD, et al. Plasma and dietary carotenoids and vitamins A, C and E and risk of colon and rectal cancer in the European Prospective Investigation into Cancer and Nutrition. Int J Cancer. 2014;135(12):2930-2939. (PubMed)

401. Leermakers ET, Darweesh SK, Baena CP, et al. The effects of lutein on cardiometabolic health across the life course: a systematic review and meta-analysis. Am J Clin Nutr. 2016;103(2):481-494. (PubMed)

402. Lees AM, Mok HYI, Lees RS, McLuskey A & Grundy SM (1977) Plant sterols as cholesterol-lowering agents: clinical trials in patients with hypercholesterolemia and studies of sterol balance. Atherosclerosis 28, 325–338.

403. Lees RS & Lees AM (1976) Effects of sitosterol therapy on plasma lipid and lipoprotein concentrations. In Lipoprotein Metabolism, pp. 119–124 [H Greten, editor]. Berlin, Heidelberg and New York: Springer.

404. Lehtonen, H.-M., Suomela, J.-P., Tahvonen, R., Yang, B., Venojarvi, M., Viikari, J., et al. (2011). Different berries and berry fractions have various but slightly positive effects on the associated variables of metabolic diseases on overweight and obese women. European Journal of Clinical Nutrition, 65(3), 394e401.

405. Leighton P (2002) Selling wellness leads to greener pastures. Functional Ingredients November.

406. Levenstein H (1988) Revolution at the Table: 3e Transformation of the American Diet. New York: Oxford University Press.

407. Leoncini E, Nedovic D, Panic N, Pastorino R, Edefonti V, Boccia S.

Carotenoid intake from natural sources and head and neck cancer: a systematic review and meta-analysis of epidemiological studies. Cancer Epidemiol Biomarkers Prev. 2015;24(7):1003-1011. (PubMed)

408. Leonel AJ, Alvarez-Leite JI (2012) Butyrate: implications for intestinal function. Curr Opin Clin Nutr Metab Care 15: 474-479.
409. Leppala JM, Virtamo J, Fogelholm R, Huttonen JK,Albanes D, Taylor PR. Controlled trial of alpha-tocopherol and beta-carotene supplements on stroke incidence and mortality in male smokers. Arterioscler Thromb Vasc Biol. 2000;20(1):230-235
410. Letenneur L, Proust-Lima C, Le Gouge A, Dartigues JF, Barberger-Gateau P. (2007). Flavonoid intake and cognitive decline over a 10-year period. American Journal of Epidemiology 15;165(12):1364-71.
411. Levenstein H (1993) Paradox of Plenty: A Social History of Eating in Modern America. New York: Oxford University Press.
412. Lin, J.-K., & Lin-Shiau, S.-Y. (2006). Mechanisms of hypolipidemic and anti-obesity effects of tea and tea polyphenols. Molecular Nutrition and Food Reserach, 50(2), 211e217.
413. Lin, M.-C., Hsu, P.-C., & Yin, M.-C. (2013). Protective effects of Houttuynia cordata aqueous extract in mice consuming a high saturated fat diet. Food Functions, 4(2), 322e327.
414. Ling WH & Jones PJH (1995) Minireview. Dietary phytosterols: a review of metabolism, benefits and side effects. Life Sciences 57, 195–206.
415. Liu R, Wang T, Zhang B, et al. Lutein and zeaxanthin supplementation and association with visual function in age-related macular degeneration. Invest Ophthalmol Vis Sci. 2015;56(1):252-258. (PubMed)
416. Liu, X.Q. and Li, Y.H. 2000. Epidemiological and nutritional research on prevention of cardiovascular disease in China. British Journal of Nutrition 84, Supplement, 2, S199-S203.
417. Lock, A.L. and Baumann, D.E. 2004. Modifying milk fat composition of dairy cows to enhance fatty acids beneficial to human health. Lipids, 39(12), 1197-1206

418. Loizzo, M. R., Lecce, G. D., Boselli, E., Menichini, F., & Frega, N. G. (2011). Inhibitory activity of phenolic compounds from extra virgin olive oils on the enzymes involved in diabetes, obesity and hypertension. Journal of Food Biochemistry, 35(2), 381e399.

419. Lopez-Miranda, J., Perez-Jimenez, F., Ros, E., De Caterina, R., Badimon, L., Covas, M. I., et al. (2010).

420. Olive oil and health: Summary of the II international conference on olive oil and health consensus report, (Spain) 2008. Nutrition, Metabolism and Cardiovascular Disease, 20(4), 284e294.

421. Lowik MR, van Dokkum W, Kistemaker C, Schaafsma G & Ockhuizen T (1993) Body composition, health status and urinary magnesium excretion among elderly people (Dutch Nutrition Surveillance System). Magnes Res 6, 223–232.

422. Lupton, J. R. Scientific substantiation of claims in the USA: focus on functional foods. European Journal of Nutrition 2009, 48: S27–S31

423. MacLatchy DL & Van der Kraak GJ (1995) The phytoestrogen beta-sitosterol alters the reproductive endocrine status of coldfish. Toxicology and Applied Pharmacology 134, 305–312

424. Maki KC, Shinnick F, Seeley MA, Veith PE, Quinn LC, Hallissey PJ, Temer A, Davidson MH (2003). Food products containing free tall oil-based phytosterols and oat beta-glucan lower serum total and LDL cholesterol in hypercholesterolemic adults. Journal of Nutrition 133(3):808-13

425. Malin DH, Lee DR, Goyarzu P, Chang YH, Ennis LJ, et al. (2011) Short term blueberry-enriched diet prevents and reverses object recognition memory loss in aging rats. Nutrition 27: 338-342.

426. Malini T & Vanithakumari G (1993) Effect of b-sitosterol on uterine biochemistry: A comparative study with estradiol and progesterone. Biochemistry and Molecular Biology International 31, 659–668.

427. Manhas A, Farmer JA (2004). Hypolipidemic therapy and cholesterol absorption. Curr Atheroscler Rep. 6:89–93

428. Manning B.B., Li, M.H., Robinson, E.H. and Peterson, B.C. 2006. Enrichment of channel catfish (Ictalurus punctatus) fillets with conjugated linoleic acid and omega-3 fatty acids by dietary manipulation. Aquaculture, 261(1), 337-342

429. Mark-Herbert, C. (2004). Innovation of a new product category—Functional foods. Technovation, 24, 713–719

430. Mares JA, LaRowe TL, Snodderly DM, et al. Predictors of optical density of lutein and zeaxanthin in retinas of older women in the Carotenoids in Age-Related Eye Disease Study, an ancillary study of the Women's Health Initiative. Am J Clin Nutr. 2006;84(5):1107-1122. (PubMed)

431. Mares-Perlman JA, Brady WE, Klein R, et al. Serum antioxidants and age-related macular degeneration in a population-based case-control study. Arch Ophthalmol. 1995;113(12):1518-1523. (PubMed)

432. Mares-Perlman JA, Fisher AI, Klein R, et al. Lutein and zeaxanthin in the diet and serum and their relation to age-related maculopathy in the third national health and nutrition examination survey. Am J Epidemiol. 2001;153(5):424-432. (PubMed)

433. Mares-Perlman JA, Klein R, Klein BE, et al. Association of zinc and antioxidant nutrients with age-related maculopathy. Arch Ophthalmol. 1996;114(8):991-997. (PubMed)

434. Mares-Perlman JA, Millen AE, Ficek TL, Hankinson SE. The body of evidence to support a protective role for lutein and zeaxanthin in delaying chronic disease. Overview. J Nutr. 2002;132(3):518S-524S. (PubMed)

435. Marion Bliss, R. (2014). Delicious, nutritious, and a colourful dish for the holidays. Agricultural Research, [online] Volume 62(10), p. 14-15. Available from: https://agresearchmag.ars.usda.gov/AR/archive/2014/Nov/potatoes1114.pdf [Accessed 24 January 2017].

436. Maskarinec, G., Oum, R., Chaptman, A. K., & Ognjanovic, S. (2009). Inflammatory markers in a randomised soya intervention among men. British Journal of Nutrition, 101(12), 1740e1744.

437. Matilsky DK, Ndekha M, Manary MJ (2009) Supplementary feeding with fortified spreads results in higher recovery rates than with a corn/soy blend in moderately wasted children. Journal of Nutrition 139(4):773–8

438. Matsufuji, H., Matsui, T., Seki, E., Osajima, K., Nakashima, M., Osajima, Y. And Matsufuji, H. 1994.

439. Angiotensin I-converting enzyme inhibitory peptides in an alkaline protease hydrolysate derived from sardine muscle. Bioscience Biotechnology and Biochemistry, 58(12), 2244-2245

440. Matsui, T., Matsufuji, H., Seki, E., Osajima, K., Nakashima, M. and Osajima, Y. 1993. Inhibition of angiotensin I-converting enzyme by Bacillus licheniformis alkaline protease hydrolysates derived from 29 sardine muscle. Bioscience Biotechnology and Biochemistry, 57(6), 922-925.

441. Mattson FH, Grundy SM, Crouse JR: Optimizing the effect of plant sterols on cholesterol absorption in man. Am J Clin Nutr. 1982, 35: 697-700.

442. Matvienko OA, Lewis DS, Swanson M, Arndt B, Rainwater DL, Stewart J, Alekel DL (2002). A single daily dose of soybean phytosterols in ground beef decreases serum total cholesterol and LDL cholesterol in young, mildly hypercholesterolemic men. American Journal of Clinical Nutrition 76(1):57-64

443. McClements, D.J. and Decker, E.A. (eds). 2009. Designing Functional Foods: - Measuring and controlling food structure breakdown and nutrient absorption. CRC Press, Boca Raton, Florida, USA

444. McCue, P., Kwon, Y.-I., & Shetty, K. (2005). Anti-diabetic and antihypertensive potential of sprouted and solid-state bioprocessed soybean. Asia Pacific Journal of Clinical Nutrition, 14(2), 145e152.

445. McCance R Widdowson EM (1956) Breads, White and Brown; 3eir Place in 3ought and Social History. Philadelphia: Lippincot.

446. McCollum EV (1957). A History of Nutrition. The Riverside Press: Cambridge, MA.

447. McCue, P., Vattem, D., & Shetty, K. (2004). Inhibitory effect of clonal oregano extracts against porcine pancreatic amylase in vitro. Asia Pacific Journal of Clinical Nutrition, 13(4), 401e408.

448. McMichael, A.J. (2000). The urban environment and health in a world of increasing globalization: issues for developing countries. Bull World Health Organ, 78, 1117-1126.

449. McNeil NI (1984) He contribution of the large intestine to energy supplies in man. Am J Clin Nutr 39: 338-342.

450. McQuillan BM, Hung J, Beilby JP, Nidorf M, Thompson PL. Antioxidant vitamins and the risk of carotid atherosclerosis. The Perth Carotid Ultrasound Disease Assessment study (CUDAS). J Am Coll Cardiol. 2001;38(7):1788-1794. (PubMed)

451. Medjakovic, S., & Jungbauer, A. (2013). Pomegranate: a fruit that ameliorates metabolic syndrome. Food Functions, 4(1), 19e39.

452. Melendez-Martinez AJ, Mapelli-Brahm P, Benitez-Gonzalez A, Stinco CM. A comprehensive review on the colorless carotenoids phytoene and phytofluene. Arch Biochem Biophys. 2015;572:188-200. (PubMed)

453. Menrad, K. (2003). Market and marketing of functional food in Europe. Journal of Food Engineering, 56, 181–188

454. Mensink RP, Ebbing S, Lindhout M, Plat J, van Heugten MM (2002). Effects of plant stanol esters supplied in low-fat yoghurt on serum lipids and lipoproteins, non-cholesterol sterols and fat soluble antioxidant concentrations. Atherosclerosis 160(1):205-13

455. Mensink RP, Ebbing S, Lindhout M, Plat J, van Heugten MM (2002). Effects of plant stanol esters supplied in low-fat yoghurt on serum lipids and lipoproteins, non-cholesterol sterols and fat soluble antioxidant concentrations. Atherosclerosis. 160(1):205-13

456. Messina, M. and Barnes, S. (1991). The role of soy products in reducing risk of cancer. Journal of the National Cancer Institute 83, p. 541-546.

457. Metchnikof E (1907) The prolongation of life. Optimistic studies. London: Butterworth-Heinemann

458. MHLW (2017). Food for Specified Health Uses (FOSHU). Available from: http://www.mhlw.go.jp/english/topics/foodsafety/fhc/02.html [Accessed 24 January 2017].

459. Michnovicz, JJ. and Bradlow, HL. (1991). Altered estrogen metabolism and excretion in humans following consumption of indole carbinol. Nutrition and Cancer 16, p. 59-66.

460. Micinski, J., Zwierzchowski, G., Kowalski, I. M., Szarek, J., Pierozynski, B., & Raistenskis, J. (2012). The effects of bovine milk fat on human health. Polish Annals of Medicine, 19(2), 170e175. microbiome project. Nature 449:804–810.

461. Miettinen TA, Tilvis RS, Kesaniemi YA. Serum plant sterols and cholesterol precursors reflect cholesterol absorption and synthesis in volunteers of a randomly selected male population. Am J Epidemiol. 1990;131:20–31

462. Miettinen TA, Vanhanen H: Dietary sitostanol related to absorption, synthesis and serum level of cholesterol in different apolipoprotein E phenotypes. Atherosclerosis. 1994, 105: 217-226.

463. Miettinen TA. Phytosterols – what plant breeders should focus on. J Sci Food Agric. 2001; 81: 895- 903.

464. Minamisawa M, Yoshida S & Takai N (2004) Determination of biologically active substances in roasted coffees using a diode-array HPLC system. Anal Sci 20, 325–328.

465. Miniello VL, Moro GE, Tarantino M, Natile M, Granieri L, Armenio L (2003). Soy-based formulas and phyto-oestrogens: a safety profile. Acta Paediatr Suppl. 91(441):93-100

466. Ministry of Health, Labour, and Welfare, Japan. Foods for Specified Health Uses (FOSHU)

467. [Accessed 22.02.17]. Available from: http://www.mhlw.go.jp/english/topics/foodsafety/fhc/02.html

468. Mintz S (1996) Food and its relationship to concepts of power. In: Tasting Food, Tasting Freedom: Excursions Into Eating, Culture, and the Past, pp. 17–32. Boston: Beacon Press.

469. Miranda, J., Fernandez-Quintela, A., Macarulla, M. T., Churruca, I., Garcia, C., Rodriguez, V. M., et al. (2009). A comparison between conjugated linolenic acid and conjugated linoleic acid effects on body fat, serum parameters and liver composition. Journal of Physiology and Biochemistry, 65(1), 25e32.

470. Mitchell JH., Cawood E., Kinniburgh D., Provan A., Collins AR., Irvine DS (2001). Effect of a phytoestrogen food supplement on reproductive health in normal males. Clinical Science 100(6) pp 613-618

471. Moayyedi P, Ford AC, Talley NJ: The efficacy of probiotics in the treatment of irritable bowel syndrome: a systematic review. Gut. 2010, 59: 325-332.

472. Moghadasian MH, McManus BM, Godin DV, Rodrigues B,

Frohlich JJ. Proatherogenic and antiatherogenic effects of probucol and phytosterols in apolipoprotein E-deficient mice: possible mechanisms of action. Circulation. 1999;99:1733–1739.

473. Moghadasian MH, McManus BM, Pritchard PH, Frohlich JJ. "Tall oil"-derived phytosterols reduce atherosclerosis in ApoE-deficient mice. Arterioscler Thromb Vasc Biol. 1997;17:119–126.

474. Moghe, S. S., Juma, S., Imrhan, V., & Vijayagopal, P. (2012). Effect of blueberry polyphenols on 3T3-F442A preadipocyte differentiation. Journal of Medicinal Food, 15(5), 448e452.

475. Mohamed, S., Hashim, S. N., & Abdul-Rahman, H. (2012). Seaweeds: a sustainable functal food for complementary and alternative therapy. Trends in Food Science and Technology, 23,(2), 83e96.

476. Moher D, Liberati A, Tetzlaff J, Altman DG; The PRISMA Group. Preferred Reporting Items for Systematic Reviews and MetaAnalyses: The PRISMA Statement. Ann Intern Med. 2009;151(4):264-269

477. Mollet, B., and Rowland, I. (2002). Functional foods: At the frontier between food and pharma. Current Opinion in Biotechnology, 13, 483–485.

478. Montrose DC, Floch MH: Probiotics used in human studies. J Clin Gastroenterol. 2005, 39 (6): 469-484

479. Moreau R, Whitaker B, Hicks K. Phytosterols, phytostanols, and their conjugates in foods: structural diversity, quantitative analysis, and health-promoting uses. Prog Lipid Res. 2002;41:457.

480. Morris MC, Evans DA, Tangney CC, Bienias JL, Wilson RS. Associations of vegetable and fruit consumption with age-related cognitive change. Neurology. 2006;67(8):1370-1376. (PubMed)

481. Morris, M.C., Sacks, F. and Rosner, B. 1993. Does fish oil lower blood pressure? A metaanalysis of controlled trials. Circulation, 88(2), 523-533.

482. Moyer, V. A. U S Preventive Services Task Force. Vitamin, mineral, and multivitamin supplements for the primary prevention of cardiovascular disease and cancer: U.S. Preventive services Task Force recommendation statement. Ann Intern Med. 2014;160(8):558-564. (PubMed)

483. Mueller, M., & Jungbauer, A. (2009). Culinary plants, herbs and spices e a rich source of PPARg ligands. Food Chemistry, 117(4), 660e667.

484. Muir, A.D. 2005. Natural peptides in blood pressure control. Agro Food Industry Hi-Technology, 16(5), 15-17.

485. Mulry, M.C. (2002). Functional foods – past, present and future. Natural Foods Merchandiser, 23(3), 82

486. Mumford SL, Kim S, Chen Z, Boyd Barr D, Buck Louis GM. (2015). Urinary Phytoestrogens Are Associated with Subtle Indicators of Semen Quality among Male Partners of Couples Desiring Pregnancy. Journal of Nutrition 145(11):2535-41.

487. Murase, T., Misawa, K., Minegishi, Y., Aoki, M., Ominami, H., Suzuki, Y., et al. (2011). Coffee polyphenols suppress diet-induced body fat accumulation by downregulating SREBP-1c and related molecules in C57BL/6J mice. American Journal of Physiology e Endocrinology and Metabolism, 300(1), E122eE133.

488. Murase, T., Yokoi, Y., Misawa, K., Ominami, H., Suzuki, Y., Shibuya, Y., et al. (2012). Coffee polyphenols modulate wholebody substrate oxidation and suppress postprandial hyperglycaemia, hyperinsulinaemia and hyperlipidaemia. British Journal of Nutrition, 107(12), 1757e1765.

489. Murata, Y., Kaneniwa, M. and Yamashita, Y. 1998. Composition of free amino acids and related compounds in the edible portion of ten salmonid species. Bulletin of the National Research Institute of Fisheries Science (Japan), 11, 65-73.

490. Murgas Torrazza, R. & Neu, J. (2011). The developing intestinal microbiome and its relationship to health and disease in the neonate. Journal of Perinatology 31, S29-S34. doi:10.1038/jp.2010.172

491. Murphy, M.G., Wright, V., Scott, J., Timmons, A. and Ackman, R.G. 1999. Dietary menhayden, seal, and corn oils differentially affect lipid and ex vive eicosanoid and thiobarbituric acid reactive substances generation in guinea pig. Lipids, 34, 115-124

492. Muzzarelli, R.A. 1996. Chitosan-based dietary foods. Carbohydrate Polymers, 29, 309-316. Ness, A.R., Hughes, J., Elwood, P.C., Whitley, E., Smith, J.D. and Burr, M.L. 2002. The long-term effect of dietary advice in men with coronary disease: follow-up of the

Diet and Reinfarction Trial (DART). European Journal of Clinical Nutrition, 56(6), 512-518.

493. N de Jong, A Zuur, M C J Wolfs, G C W Wendel-Vos, J M A van Raaij and A J Schuit (2007). Exposure and effectiveness of phytosterol/-stanol-enriched margarines. Eur J Clin Nutr 61 (12), 1407-1415

494. Nagourney, RA. (1998) Garlic: Medicinal food or nutritious medicine? Journal of Medicinal Food 1, p. 13-28.

495. Nakachi, K., Suemasu, K., Suga, K., Takeo, T., Imai, K., and Higashi, Y. (1998). Influence of drinking green tea on breast cancer malignancy among Japanese patients. Japanese Journal of Cancer Research 89, p. 254-261.

496. Nakazato, K., Song, H., & Waga, T. (2006). Effects of dietary apple polyphenol on adipose tissues weights in Wistar rats. Experimental Animals, 55(4), 383e389. 126

497. Narita, Y., & Inouye, K. (2009). Kinetic analysis and mechanism on the inhibition of chlorogenic acid and its components against porcine pancreas a-amylase isozymes I and II. Journal of Agricultural and Food Chemistry, 57(19), 9218e9225.

498. Narkhede, M. B. (2012). Evaluation of alpha amylase inhibitory potential of four traditional culinary leaves. Asian Journal of Pharmaceutical and Clinical Research, 5(Suppl. 2), 75e76. Nature. 464:59–65.

499. Nestel P, Cehun M, Pomeroy S, Abbey M, Weldon G (2001). Cholesterol-lowering effects of plant sterol esters and non-esterified stanols in margarine, butter and low-fat foods. European Journal of Clinical Nutrition 55(12):1084-90

500. Nestel P, Cehun M, Pomeroy S, Abbey M, Weldon G (2001). Cholesterol-lowering effects of plant sterol esters and non-esterified stanols in margarine, butter and low-fat foods. European Journal of Clinical Nutrition 55(12):1084-90

501. Nestle, M. (1998). Broccoli sprouts in cancer prevention. Nutrition Reviews 56, p. 127-130.

502. Nestle, PJ., Yamashita, T., Sasahara, T., Pomeroy, S., Dart, A., Komesaroff, P., Owen, A., and Abbey, M. (1997). Soy isoflavones improve systemic arterial compliance but not plasma lipids in menopausal and perimenopausal women. Arteriosclerosis,

Thrombosis and Vascular Biology 17, p. 3392-3398.

503. Nestle M (2002) Food Politics: How the Food Industry Influences Nutrition and Health. Berkeley and Los Angeles: University of California Press.

504. Nguyun TT (1999). The cholesterol-lowering action of plant stanol esters. J Nutr 129, 2109–2112.

505. Nicoli MC, Anese M, Manzocco L & Lerici CR (1997) Antioxidant properties of coffee brews in relation to the roasting degree. Lebens Wiss Technol 30, 292–297.

506. Nicoli MC, Anese M & Parpinel M (1999) Influence of processing on the antioxidant properties of fruit and vegetables. Trends Food Sci Technol 10, 94–100.

507. Nieman, D. C., Cayea, E. J., Austin, M. D., Henson, D. A., McAnulty, S. R., & Jin, F. (2009). Chia seed does not promote weight loss or alter disease risk factors in overweight adults. Nutrition Research, 29(6), 414e418.

508. Niemann, B. 2010. Functional ingredients: how much should we add to foods? In: Gormley, T.R. and Holm, F. (eds) Functional Foods: Some Pointers for Success. UCD Institute of Food and Health, University College Dublin, Dublin, Ireland, http://www.ucd.ie/t4cms/ffnet%20funct%20fds%20final%20epublication%20jan%202011.pdf, 66-69

509. Nikolakopoulou, Z., Nteliopoulos, G., Michael-Titus, A.T. and Parkinson, K. 2013. Omega-3 polyunsaturated fatty acids selectively inhibit growth in neoplastic oral keratinocytes by differentially activating ERK1/2. Carcinogenesis, doi:10.1093?carcin/bgt257

510. Nimsate K, Mohamed A, Jianmei Y. (2010). Development of a fortified peanut-based infant formula of severly malnourished children. Int J Food Sci Technol. 45(10):1965–1972.

511. Niva M (2007) All foods affect health: Understandings of functional foods and healthy eating among health-oriented Finns. Appetite 48(3) 384–393.

512. Noguchi, M. 2003. Induction of hepatocyte growth factor and inhibition of proliferation by extracellular matrix component derived from fish skin of Reinhararditus hippoglossoides. Fisheries Science (Japan), 69(2), 401-407

513. Noone, E.J., Roche, H.M., Nugent, A.P. and Gibney, M.J. 2002. The effect of dietary supplementation using isomeric blends of conjugated linoleic acid on lipid metabolism in healthy subjects. British Journal of Nutrition 88, 243-251

514. Nys, M. and Debruyne, I. 2011. Lipids and brain 2: a symposium on lipids and brain health. Inform, 22 (7), 397-399.

515. O'Leary, D. 2010. Calcium: Filling the nutritional gap means eating well - - supplement naturally and 'BONE APPETIT'. In: Gormley, T.R. and Holm, F. (eds) Functional Foods: Some Pointers for Success.

516. UCD Institute of Food and Health, University College Dublin, Dublin 4, Ireland, http://www.ucd.ie/t4cms/ffnet%20funct%20fds%20final%20epublication%20jan%202011.pdfe, 85-88

517. Obiro, W. C., Zhang, T., & Jiang, B. (2009). Starch blocking stability of the Phaseolus vulgaris alpha-amylase inhibitor (a-AI1). American Journal of Food Technology, 4(1), 9e19.

518. Ofek, I., Goldhar, J., Zafriri, D., Lis, H., Adar, R., and Sharon, N. (1991). Anti-Escherichia coli adhesin activity of cranberry and blueberry juices. New England Journal of Medicine 324, p. 1599.

519. Ogawa, M., Portier, R.J., Moody, M.W., Bell, J., Schexnayder, M.A. and Losso, J.N. 2004. Biochemical properties of bone and scale collagens isolated from the subtropical black drum (Pogonia cromis) and sheepshead seabass (Archosargus probatocephalus). Food Chemistry, 88(4), 495-501

520. Ogunmodede, O. S., Saalu, L. C., Ogunlade, B., Akunna, G. G., & Oyewopo, A. O. (2012). An evaluation of the hypoglycemic, antioxidant and hepatoprotective potentials of onion (Allium cepa L.) on alloxan-induced diabetic rabbits. International Journal of Pharmacology, 8(1), 21e29.

521. Ohama H, Ikeda H, Moriyama H (2006). Health foods and foods with health claims. Toxicology 221, 95–111.

522. Ohba, R., Deguchi, T., Kishikawa, M., Arysad, F., Morimura, S. and Kida, K. 2003. Physiological functions of enzymatic hydrolysates of collagen or keratin contained in livestock and fish waste. Food Science and Technology Research, 9(1), 91-93

523. Oksanen PJ, Salminen S, Saxelin M, et al. Prevention of travellers'

diarrhoea by Lactobacillus GG. Ann Med 1990;22:53–6.

524. Olsen, H., Aaby, K., & Borge, G. I. A. (2009). Characterization and quantification of flavonoids and hydroxycinnamic acids in curly kale (Brassica oleracea L. convar. acephala var. sabellica) by HPLC-DAD-ESI-MSn. Journal of Agricultural and Food Chemistry, 57(7), 2816e2825.

525. Omenn GS, Goodman GE, Thornquist MD, et al. Effects of a combination of β-carotene and vitamin A on lung cancer and cardiovascular disease. N Engl J Med. 1996;334(18):1150-1155.

526. Omenn GS, Goodman GE, Thornquist MD, et al. Risk factors for lung cancer and for intervention effects in CARET, the Beta-Carotene and Retinol Efficacy Trial. J Natl Cancer Inst. 1996;88(21):1550-1559. (PubMed)

527. Orphan Drug Act." Orphan Drug Act. Web. 19 Mar. 2015

528. Osganian SK, Stampfer MJ, Rimm E, Spiegelman D, Manson JE, Willett WC. Dietary carotenoids and risk of coronary artery disease in women. Am J Clin Nutr. 2003;77(6):1390-1399. (PubMed)

529. Ostlund RE Jr, Spilburg CA, Stenson WF (1999). Sitostanol administered in lecithin micelles potently reduces cholesterol absorption in humans. American Journal of Clinical Nutrition 70(5):826-31

530. Ostlund RE, Spilburg CA, Stenson WF. Sitostanol administered in lecithin micelles potently reduces cholesterol absorption in humans. Am J Clin Nutr. 1999;70:826–831

531. Ostlund RE, Spilburg CA, Stenson WF: Sitostanol administered in lecithin micelles potently reduces cholesterol absorption in humans. Am J Clin Nutr. 1999, 70: 826-831.

532. O'Sullivan GC, Kelly P, O'Halloran S, Collins C, Collins JK, Dunne C, Shanahan F (2005). Probiotics: an emerging therapy. Curr Pharm Des. 2005; 11(1):3-10.

533. Ozaki K, Okamoto M, Fukasawa K, et al. Daily intake of β-cryptoxanthin prevents bone loss by preferential disturbance of osteoclastic activation in ovariectomized mice. J Pharmacol Sci. 2015;129(1):72-77. (PubMed)

534. Panchal, S. K., Poudyal, H., Waanders, J., & Brown, L. (2012). Coffee

extract attenuates changes in cardiovascular and hepatic structure and function without decreasing obesity in high-carbohydrate, high-fat diet-fed male rats. Journal of Nutrition, 142(4), 690e697.

535. Pandya, U., Chandra, A., Awasthi, S., Jin, G. F., Piper, J. T., Godley, B. F., et al. (2000). Attenuation of galactose cataract by low levels of dietary Curcumin. Nutrition Research, 20(4), 515e526. Patel, D. K.,

536. Prasad, S. K., Kumar, R., & Hemalatha, S. (2011). Cataract, A major secondary complication of diabetes, its epidemiol and an overview on major medicinal plants screened for anticataract activity. Asian Pacific Journal of Tropical Disease, 1(4), 323e329.

537. Panic N, Nedovic D, Pastorino R, Boccia S, Leoncini E. Carotenoid intake from natural sources and colorectal cancer: a systematic review and meta-analysis of epidemiological studies. Eur J Cancer Prev. 2016. [Epub ahead of print] (PubMed)

538. Pariza, M.W., Park, Y. and Cook, M.E. 2001. The biologically active isomers of conjugated linoleic acid. Progress in Lipid Research, 40, 283-298

539. Parker RA, Pearce BC, Clark RW, Gordon DA, Wright JJK. Tocotrienols regulate cholesterol production in mammalian cells by post-transcriptional suppression of 3-hydroxy-3-methylglutaryl-coenzyme A reductase. J Biol Chem. 1993; 268:11230-8.

540. Patisaul, HB. And Jefferson, W. (2010). The pros and cons of phytoestrogens. Front Neuroendocrinology 31(4): 400–419

541. Patterson, A.C. and Stark, K.D. 2008. Direct determination of the fatty acid composition of daily dietary intakes incorporating nutraceuticals and functional food strategies to increase n-3 highly unsaturated fatty acids. Journal of the American College of Nutrition, 27(5), 538-546

542. Pearce JL, Hamilton JR. Controlled trial of orally administered lactobacilli in acute infantile diarrhea. J Pediatr 1974;84:261–2.

543. Pei R, Martin DA, DiMarco DM, Bolling BW (2015) Evidence for the effects of yogurt on gut health and obesity. Critical Reviews in Food Science and Nutrition

544. Pelletier X, Belbraouet S, Mirabel D, Mordret F, Perrin JL, Pages X & Derby G (1995) A diet moderate enriched in phytosterols lowers plasma cholesterol concentrations in normocholesterolemic

humans. Annals of Nutrition and Metabolism 39, 291–295.

545. Pelletier X, Belbraouet S, Mirabel D, Mordret F, Perrin JL, Pages X, Debry G: A diet moderately enriched in phytosterols lowers plasma cholesterol concentrations in normocholesterolemic humans. Ann Nutr etab. 1995, 39: 291-295.

546. Peng, C.-H., Liu, L.-K., Chuang, C.-M., Chyau, C.-C., Huang, C.-N., & Wang, C.-J. (2011). Mulberry water extracts possess an antiobesity effect and ability to inhibit hepatic lipogenesis and promote lipolysis. Journal of Agricultural and Food Chemistry, 59(6), 2663e2671.

547. Perez-Herrera, A., Delgado-Lista, J., Torres-Sanchez, L. A., RangelZuniga, O. A., Camargo, A., Moreno-Navarrete, J. M., et al. (2012). The postprandial inflammatory response after ingestion of heated oils in obese persons is reduced by the presence of phenol compounds. Molecular Nutrition and Food Research, 56(3), 510e514.

548. Perez-Torres, I., Ruiz-Ramirez, A., Banos, G., & El-Hafidi, M. (2013). Hibiscus Sabdariffa Linnaeus (Malvaceae), curcumin and resveratrol as alternative medicinal agents against metabolic syndrome. Cardiovascular and Hematology Agents in Medicinal Chemistry, 11(1), 25e37.

549. Peterson J, Garges S, Giovanni M, McInnes P, Wang L, Schloss JA, Bonazzi V, McEwen JE,

550. Picone, O., Marszalek, A., Servely, J.L. and Chavatte-Palmer, P. 2009. Effects of omega-3 supplementation in pregnant women. Journal de Gynecologie Obstetrique et Biologie de la Reproduction, 38(2), 117-124

551. Pihlanto, A., Akkanen, S., & Korhonen, H. J. (2008). ACE-inhibitory and antioxidant properties of potato (Solanum tuberosum). Food Chemistry, 109(1), 104e112.

552. Piironen GW, Price KR, Rhodes MJC, Willianson G. Tocopherols and tocotrienols in Finnish foods: Vegetables, fruits and berries. J Agric Food Chem. 1986; 34: 742-6.

553. Piironen V, Lindsay DG, Miettinen TA, Toivo J, Lampi A-M. Plant sterols: biosynthesis, biological function and their importance to human nutrition. J Sci Food Agric. 2000;80:939–966.

554. Plat J, Mensink RP (2001). Effects of plant sterols ands stanols on lipid metabolism and cardiovascular risk. Nutr Metab Cardiovasc Dis 11, 31–40.

555. Pollan M (2007) Unhappy meals. New York Times January 28. Pollan M (2008) In Defense of Food: An Eater's Manifesto. New York: Penguin Press.

556. Popkin, B.M. (2004). The Nutrition Transition: An Overview of World Patterns of Change. Nutrition Reviews, 62(7), S140-S143

557. Potter, SM. (1998). Soy protein and cardiovascular disease: The impact of bioactive components in soy. Nutrition Reviews 56(8), p. 231-235.

558. Pouteau E, Monnard I, Piguet-Welsch C, Groux MJA, Sagalowicz L, Berger A: Non-esterified plant sterols solubilized in low fat milks inhibit cholesterol absorption: a stable isotope double-blind crossover study. Eur J Nutr. 2003, 42: 154-164.

559. Pouteau EB, Monnard IE, Piguet-Welsch C, Groux MJ, Sagalowicz L, Berger A (2003). Non-esterified plant sterols solubilized in low fat milks inhibit cholesterol absorption--a stable isotope double-blind crossover study. European Journal of Nutrition 42(3):154-64

560. Pouteau EB, Monnard IE, Piguet-Welsch C, Groux MJ, Sagalowicz L, Berger A (2003). Non-esterified plant sterols solubilized in low fat milks inhibit cholesterol absorption--a stable isotope double-blind crossover study. European Journal of Nutrition 42(3):154-64.

561. Pouteau EB, Monnard IE, Piguet-Welsch C, Groux MJ, Sagalowicz L, Berger A (2003). Non-esterified plant sterols solubilized in low fat milks inhibit cholesterol absorption--a stable isotope double-blind crossover study. European Journal of Nutrition 42(3):154-64

562. Premkumar, S (2014). Fascinating Facts about Phytonutrients in Spices and Healthy Food: Scientifically Proven Facts. Xilibris editors.

563. Prior, R. L. E., Wilkes, S. R., Rogers, T., Khanal, R. C., Wu, X., et al. (2010b). Purified blueberry anthocyanins and blueberry juice alter development of obesity in mice fed an obesogenic high-fat diet. Journal of Agricultural and Food Chemistry, 58(7), 3970e3976.

564. Prior, R. L., Wilkes, S., Rogers, T., Khanal, R. C., Wu, X., Hager, T. J., et al. (2010a). Dietary black raspberry anthocyanins do not alter

development of obesity in mice fed an obesogenic high-fat diet. Journal of Agricultural and Food Chemistry, 58(7), 3977e3983.

565. Priyadarshani AM. A review on factors influencing bioaccessibility and bioefficacy of carotenoids. Crit Rev Food Sci Nutr. 2015. [Epub ahead of print].
566. Promoting Health by Linking Agriculture, Food, and Nutrition Holm L (2003) Food health policies and ethics: Lay perspectives on functional foods. Journal of Agricultural and Environmental Ethics 16 531–544
567. Prosky L. (2000). When is dietary fiber considered a functional food? Biofactors 12 (1-4): 289-297
568. Puri, V. N. (1999). Banana and hypertension. Current Science, 76(8), 1064e1069.
569. Qin J, Li R, Raes J, Arumugam M, Burgdorf KS, Manichanh C, Nielsen T, Pons N, Levenez F,Qin, B., & Anderson, R. A. (2012). An extract of chokeberry attenuates weight gain and modulates insulin, adipogenic and inflammatory signalling pathways in epididymal adipose tissue of rats fed a fructose-rich diet. British Journal of Nutrition, 108(4), 581e587.
570. Qureshi AA, Pearce BC, Nor RM, Gapor A, Peterson DM, Elson CE. α-Tocopherol attenuates the impact of tocotrienol on hepatic 3-hydroxy-3-methylglutaryl coenzyme A reductase activity. J Nutr. 1996; 126:389- 94.
571. Rademacher, M., Gormley, T.R. and Walshe, T. 2004. Exploring the potential of fish bone and crab shell waste as calcium nutraceuticals. Proceedings of the 34th Annual Food Science and Technology Research Conference, University College Cork, Cork, Ireland, 39 (abstract)
572. Raghunathan, R., Naylor, R.W. and Hoyer, W.D. (2006). The Unhealthy = Tasty Intuition and Its Effects on Taste Inferences, Enjoyment and Choice of Food Products. Journal of Marketing, 70, 170-184.
573. Rai, M., & Acharya, K. (2012). Proximate composition, free rad scavenging and NOS activation properties of Ramaria aureA. Research Journal of Pharmacy and Technology, 5(11), 1421e1427.
574. Raja B.R. and Arunachalam K.D. (2011). Market potential for

probiotic nutritional supplements in India, African Journal of Business Management, 5(14), 5418-5423.

575. Ramos, A., Bandarra, N.M., Rema, P., Vaz-Pires, P., Nunes, M.L., Andrade, A.M., Cordeiro, A.R. and Valente, L.M.P. 2008. Time course deposition of conjugated linoleic acid in market-size rainbow trout (Oncorhynchus mykiss) muscle. Aquaculture, 274(2-4), 366-374

576. Rana, S.K. and Sanders, T.A. 1986. Taurine concentration in the diet, plasma, urine and breast milk of vegans compared with omnivores. British Journal of Nutrition, 56, 17-27.

577. Rebholz, C. M., Friedman, E. E., Powers, L. J., Arroyave, W. D., He, J., & Kelly, T. N. (2012). Dietary protein intake and blood pressure: a meta-analysis of randomized controlled trials. American Journal of Epidemiology, 176(Suppl. 7), S27eS43.

578. Renzi LM, Dengler MJ, Puente A, Miller LS, Hammond BR, Jr. Relationships between macular pigment optical density and cognitive function in unimpaired and mildly cognitively impaired older adults. Neurobiol Aging. 2014;35(7):1695-1699. (PubMed)

579. Reuter HD., Koch, HP, and Lawson, LD. (1996). Therapeutic effects and applications of garlic and its preparations. In: Garlic. The Science and Therapeutic Application of Allium sativum L. and Related Species, 2nd Ed., ed., H.P. Koch and L.D. Lawson, Williams & Wilkins, Baltimore.

580. Richelle M, Enslen M, Hager C, Groux M, Tavazzi I, Godin JP, Berger A, Métairon S, Quaile S, Piguet C, Sagalowicz L, Green H, Fay LB: Both free and esterified plant sterols decrease the bioavailability of β-carotene and β-tocopherol, in normocholesterolemic humans. Am J Clin Nutr. 2004

581. Richelle M, Enslen M, Hager C, Groux M, Tavazzi I, Godin JP, Berger A, Métairon S, Quaile S, Piguet-Welsch C, Sagalowicz L, Green H, Fay LB (2004). Both free and esterified plant sterols reduce cholesterol absorption and the bioavailability of beta-carotene and alpha-tocopherol in normocholesterolemic humans. American Journal of Clinical Nutrition 80(1):171-7

582. Richer S, Stiles W, Statkute L, et al. Double-masked, placebo-controlled, randomized trial of lutein and antioxidant supplementation

in the intervention of atrophic age-related macular degeneration: the Veterans LAST study (Lutein Antioxidant Supplementation Trial). Optometry. 2004;75(4):216-230.

583. Rimm EB, Stampfer MJ, Ascherio A, Giovannucci E, Colditz GA, Willett WC. Vitamin E consumption and the risk of coronary heart disease in men. N Engl J Med. 1993;328(20):1450-1456. (PubMed)

584. Rissanen T, Voutilainen S, Nyyssonen K, Salonen R, Salonen JT. Low plasma lycopene concentration is associated with increased intima-media thickness of the carotid artery wall. Arterioscler Thromb Vasc Biol. 2000;20(12):2677-2681. (PubMed)

585. Rissanen TH, Voutilainen S, Nyyssonen K, et al. Low serum lycopene concentration is associated with an excess incidence of acute coronary events and stroke: the Kuopio Ischaemic Heart Disease Risk Factor Study. Br J Nutr. 2001;85(6):749-754. (PubMed)

586. Rissanen TH, Voutilainen S, Nyyssonen K, Salonen R, Kaplan GA, Salonen JT. Serum lycopene concentrations and carotid atherosclerosis: the Kuopio Ischaemic Heart Disease Risk Factor Study. Am J Clin Nutr. 2003;77(1):133-138. (PubMed)

587. Roberfroid, M.B. (2002). Global view on functional foods: European perspectives. British Journal of Nutrition, 88(2), S133–S138.

588. Roberfroid MB. Prebiotics and synbiotics: concepts and nutritional properties. Br J Nutr 1998;80:S197–202

589. Roberts, LM., McCahon, D., Holder, R., Wilson, S., and Hobbs, FDR. (2013). A Randomised Controlled Trial of a Probiotic 'Functional Food' in the Management of Irritable Bowel Syndrome. BMC Gastroenterol. 2013;13(45)

590. Robinson, C.J., Bohannan, B.J.M. & Young, V.B. (2010). From structure to function: the ecology of host-associated microbial communities. Microbiology and Molecular Biology Reviews, 74(3), 453-476. doi: 10.1128/MMBR.00014-10

591. Rojo, L. E., Ribnicky, D., Logendra, S., Poulev, A., Rojas-Silva, P., Kuhn, P., et al. (2012). In vitro and in vivo anti-diabetic effects of anthocyanins from Maqui Berry (Aristotelia chilensis). Food Chemistry, 131(2), 387e396.

592. Rooban, B. N., Lija, Y., Biju, P. G., Sasikala, V., Sahasranamam, V., & Abraham, A. (2009). Vitex negundo attenuates calpain activation

and cataractogenesis in selenite models. Experimental Eye Research, 88(3), 575e582.

593. Roopchand, D. E., Kuhn, P., Rojo, L. E., Lila, M. A., & Raskin, I. (2013). Blueberry polyphenol-enriched soybean flour reduces hyperglycemia, body weight gain and serum cholesterol in mice. Pharmacology Research, 68(1), 59e67.

594. Ros, E. (2010). Health benefits of nut consumption. Nutrients, 2(7), 652e682.

595. Rose, E.L. and Holub, B.J. 2006. Effects of a liquid egg product containing fish oil on selected cardiovascular disease risk factors: a randomized crossover trial. Food Research International, 39(8), 910-916

596. Rosenow, A., Noben, J.-P., Jocken, J., Kallendrusch, S., FischerPosovszky, P., Mariman, E. C. M., et al. (2012). Resveratrolinduced changes of the human adipocyte secretion profile. Journal of Proteome Research, 11(9), 4733e4743.

597. Saarela, M., Lähteenmäki, L., Crittenden, R., Salminen, S., & Mattila-Sandholm, T. (2002). Gut bacteria and health foods—the European perspective. International journal of food microbiology, 78(1), 99-117.

598. Saavedra J, Tschernia A, Moore N, et al. Gastrointestinal function in infants consuming a weaning food supplemented with oligofructose, a prebiotic. J Pediatr Gastroenterol Nutr 1999;29:513

599. Sahni S, Hannan MT, Blumberg J, Cupples LA, Kiel DP, Tucker KL. Inverse association of carotenoid intakes with 4-y change in bone mineral density in elderly men and women: the Framingham Osteoporosis Study. Am J Clin Nutr. 2009;89(1):416-424. (PubMed)

600. Sahni S, Hannan MT, Blumberg J, Cupples LA, Kiel DP, Tucker KL. Protective effect of total carotenoid and lycopene intake on the risk of hip fracture: a 17-year follow-up from the Framingham Osteoporosis Study. J Bone Miner Res. 2009;24(6):1086-1094. (PubMed)

601. Sahyoun NR, Jacques PF, Russell RM. Carotenoids, vitamins C and E, and mortality in an elderly population. Am J Epidemiol. 1996;144(5):501-511. (PubMed)

602. Saito, M., Ueno, M., Ogino, S., Kubo, K., Nagata, J., & Takeuchi, M.

(2005). High dose of Garcinia cambogia is effective in suppressing fat accumulation in developing male Zucker obese rats, but highly toxic to the testis. Food Chemistry and Toxicology, 43(3), 411e419.

603. Sakaguchi, M., Murata, M. and Kawai, A 1982. Taurine levels in some tissues of yellowtail (Seriola quinqueradiata) and mackerel (Scomber japonicus). Agricultural and Biological Chemistry, 46(11), 2857-2858

604. Salen G, Ahrens EH & Grundy SM (1970) Metabolism of bsitosterol in man. Journal of Clinical Investigation 49, 952–967

605. Sanchez-Alonso, L., Solas, M.T. and Borderias, A.J. 2007. Physical study of minced fish muscle with a white grape by-product added as an ingredient. Journal of Food Science, 72(2), 94-101

606. Sanders TH. Non detectable levels of trans-fatty acids in peanut butter. Journal of Agriculture and Food Chemistry 49(5):2349–2351

607. Sankhari, J. M., Thounaojam, M. C., Jadeja, R. N., Devkar, R. V., & Ramachandran, A. V. (2012).

608. Anthocyanin-rich red cabbage (Brassica oleracea L.) extract attenuates cardiac and hepatic oxidative stress in rats fed an atherogenic diet. Journal of Science and Food Agriculture, 92(8), 1688e1693.

609. Santos EE, Lauria DC & Porto da Silveira CL (2004) Assessment of daily intake of trace elements due to consumption of foodstuffs by adult inhabitants of Rio de Janeiro city. Sci Total Env 327, 69–79.

610. Saraswat, M., Suryanarayana, P., Reddy, P. Y., Patil, M. A., Balakrishna, N., & Reddy, G. B. (2010). Antiglycating potential of Zingiber officinalis and delay of diabetic cataract in rats. Molecular Vision, 16, 1525e1537.

611. Sasaki, R., Nishimura, N., Hoshino, H., Isa, Y., Kadowaki, M., Ichi, T., et al. (2007). Cyanidin 3-glucoside

612. ameliorates hyperglycemia and insulin sensitivity due to downregulation of retinol binding protein 4 expression in diabetic mice. Biochemistry and Pharmacology, 74(11), 1619e1627.

613. Schallich, E. and Gormley, T.R. 1996. Condition factor, fat content and flavour of farmed and wild salmon. Farm and Food 6(3), 28- 31.

614. Scherbakov, AM., Andreeva OE. (2015). Apigenin Inhibits Growth

of Breast Cancer Cells: The Role of ERα and HER2/neu. Acta Naturae. 7(3):133-9.

615. Schini-Kerth, V. B., Etienne-Selloum, N., Chataigneau, T., & Auger, C. (2011). Vascular protection by natural product-derived polyphenols: in vitro and in vivo evidence. Planta Medica, 77(11), 1161e1167.

616. Schlierf G, Oster P, Heuck CC, Raetzer H & Schellenberg B (1978) Sitosterolemia in juvenile type II hyperlipoproteinemia. Atherosclerosis 30, 245–248

617. Schmidt, DR. and Sobota, AE. (1988). An examination of the anti-adherence activity of cranberry juice on urinary and nonurinary bacterial isolates. Microbios 55, p. 173- 181.

618. Schram, E., Pedrero, Z., Camara, C., van der Heul, J.W. and Luten, J.B. 2008. Enrichment of African catfish with functional selenium originating from garlic. Aquaculture Research, 39(8), 850-860.

619. Scrinis G (2008) Functional foods or functionally marketed foods? A critique of and alternatives to, the category of 'functional foods.' Public Health Nutrition 11(5) 541–545.

620. Scripsema NK, Hu DN, Rosen RB. Lutein, zeaxanthin, and meso-zeaxanthin in the clinical management of eye disease. J Ophthalmol. 2015;2015:865179. (PubMed)

621. Seddon JM, Ajani UA, Sperduto RD, et al. Dietary carotenoids, vitamins A, C, and E, and advanced age-related macular degeneration. Eye Disease Case-Control Study Group. JAMA. 1994;272(18):1413-1420. (PubMed)

622. Sekirov I, Russell SL, Antunes LC, Finlay BB (2010) Gut microbiota in health and disease. Physiol Rev 90: 859-904

623. Serbinova EA, Tsuchiya M, Goth S, Kagan VE, Packer L. Antioxidant action of α-tocopherol and α-tocotrienol in membranes. In: Packer L, Fuchs J (editors). Vitamin E in health and disease. New York: Marcel Dekker; 1993. p. 235-43.

624. Sergent, T., Vanderstraeten, J., Winand, J., Beguin, P., & Schneider, Y.- J. (2012). Phenolic compounds and plant extracts as potential natural anti-obesity substances. Food Chemistry, 135(1), 68e73.

625. Sesso HD, Buring JE, Norkus EP, Gaziano JM. Plasma lycopene,

other carotenoids, and retinol and the risk of cardiovascular disease in men. Am J Clin Nutr. 2005;81(5):990-997. (PubMed)

626. Sesso HD, Buring JE, Norkus EP, Gaziano JM. Plasma lycopene, other carotenoids, and retinol and the risk of cardiovascular disease in women. Am J Clin Nutr. 2004;79(1):47-53. (PubMed)
627. Setchell, KDR., Lawson, AM, Borriello, SP, Harkness, R., Gordon, H., Morgan, DML, Kirk, DN., Adlercreutz, H., Anderson, LC., and Axelson, M. (1981). Lignan formation in man -- microbial involvement and possible roles in relation to cancer. The Lancet ii, p. 4-7.
628. Seymour, E. M., Tanone, I. I., Urcuyo-Llanes, D. E., Lewis, S. K., Kirakosyan, A., Kondoleon, M. G., et al. (2011). Blueberry intake alters skeletal muscle and adipose tissue peroxisome proliferatoractivated receptor activity and reduces insulin resistance in obese rats. Journal of Medicinal Food, 14(12), 1511e1518.
629. Sghir A, Chow JM, Mackie RI. Continuous culture selection of bifidobacteria and lactobacilli from human faecal samples using fructooligosaccharide as selective substrate. J Appl Microbiol 1998;85:769–77.
630. Shabrova, E. V., Tarnopolsky, O., Singh, A. P., Plutzky, J., Vorsa, N., & Quadro, L. (2011). Insights into the molecular mechanisms of the anti-atherogenic actions of flavonoids in normal and obese mice. PLoS ONE, 6(10), e24634.
631. Shahidi, F. 2003. Nutraceuticals and bioactives from seafood products. In: Bechtel, P. J. (ed.) Proceedings of 2nd International Seafood By-Products Conference. Alaska Sea Grant College Programme, Fairbanks, Alaska, USA, pp. 247-263.
632. Shahidi, F. 2003. Nutraceuticals and bioactives from seafood products. In: Bechtel, P. J. (ed.) Proceedings of 2nd International Seafood By-Products Conference. Alaska Sea Grant College Programme, Fairbanks, Alaska, USA, pp. 247-263.
633. Shahidi, F., Arachchi, J.K.V. and Jeon, Y-J. 1999. Food applications of chitin and chitosans. Trends in Food Science and Technology, 10, 37-51
634. Sharma, A. K., Bharti, S., Bhatia, J., Nepal, S., Malik, S., Ray, R., et al. (2012). Sesamol alleviates diet-induced cardiometabolic

syndrome in rats via up-regulating PPARg, PPAR a and e-NOS. Journal of Nutritional Biochemistry, 23(11), 1482e1489.

635. Shields DH, Corrales KM & Metallinos-Katsaras E (2004) Gourmet coffee beverage consumption among college women. J Am Diet Assoc 104, 650–653.

636. Shikov, A. N., Pozharitskaya, O. N., Makarova, M. N., Kovaleva, M. A., Laakso, I., Dorman, H. J. D., et al. (2012). Effect of Bergenia crassifolia L. extracts on weight gain and feeding behavior of rats with high-caloric diet-induced obesity. Phytomedicine, 19(14), 1250e1255.

637. Shimizu T. Newly established regulation in Japan: foods with health claims. Asia Pacific J Clin Nutr 2002;11(suppl): S94-6.

638. Shimizu T: Health claims on functional foods: the Japanese regulations and an international comparison. Nutrition Research Reviews 2003, 16: 241-252. 11.

639. Shimoda, H., Seki, E., & Aitani, M. (2006). Inhibitory effect of green coffee bean extract on fat accumulation and body weight gain in mice. BMC Complementary and Alternative Medicine, 6, 9e15.

640. Shu, XO., Zheng, Y., Cai, H., Gu, K., Chen, Z., Zheng, W., and Lu, W. (2009). Soy Food Intake and Breast Cancer Survival. Journal of the American Medical Association 302(22):2437-2443

641. Shukri, R., Mohamed, S., & Mustapha, N. M. (2010). Cloves protect the heart, liver and lens of diabetic rats. Food Chemistry, 122(4), 1116e1121.

642. Sierksma A, Weststate JA, Meijer GW. Spreads enriched with plant sterols, either esterified 4,4 dimethylsterols or free 4 desmethylsterols, and plasma total and LDL cholesterol concentrations. Br J Nutr. 1999;82:273–282.

643. Sierksma A, Weststate JA, Meijer GW: Spreads enriched with plant sterols, either esterified 4, 4 dimethylsterols or free 4 desmethylsterols, and plasma total and LDL cholesterol concentrations. Br J Nutr. 1999, 82: 273-282.

644. Sierksma, A., Weststrate, J. A., & Meijer, G. W. (1999). Spreads enriched with plant sterols, either esterified 4, 4-dimethylsterols or free 4-desmethylsterols, and plasma total-and LDL-cholesterol concentrations. British Journal of Nutrition, 82(04), 273-282.

645. Siitonen S, Vapaatalo H, Salminen S, et al. Effect of Lactobacillus GG yoghurt in prevention of antibiotic associated diarrhoea.

646. Silagy, CA. and Neil, HAW. (1994). A meta-analysis of the effect of garlic on blood pressure. Journal of Hypertension 12, p. 463-468.

647. Singer, P. and Wirth, M. 2003. Omega-3 fatty acids of marine and of vegetable origin-a state of the art. Ernahrungs- Umschau, 50(8), 296-304

648. Slanc, P., Doljak, B., Kreft, S., Lunder, M., Janes, D., & Strukelj, B. (2009). Screening of selected food and medicinal plant extracts for pancreatic lipase inhibition. Phytotherapy Research, 23(6), 874e877.

649. Slattery, C and Gormley, T.R. 2103b. Fishy business: consumer survey of fish purchasing patterns and awareness of the health properties of fish. Food Science and Technology, 27(3), 20-23

650. Slattery, C. and Gormley, T.R. 2013a. Survey on the oil content of three fish species. The Skipper, August, 30.

651. Smart EJ, De Rose RA, Farber SA. (2004). Annexin 2-caveolin 1 complex is a target of ezetimibe and regulates intestinal cholesterol transport. Proc Natl Acad Sci USA. 101:3450–3455.

652. Snellen EL, Verbeek AL, Van Den Hoogen GW, Cruysberg JR, Hoyng CB. Neovascular age-related macular degeneration and its relationship to antioxidant intake. Acta Ophthalmol Scand.

653. 2002;80(4):368-371. (PubMed)

654. Sobenin IA, Myasoedova VA, Orekhov AN (2016). Phytoestrogen-Rich Dietary Supplements in Anti-Atherosclerotic Therapy in Postmenopausal Women. Curr Pharm Des. 22(2):152-63.

655. Söderholm, P. P., Alfthan, G., Koskela, A. H., Adlercreutz, H., & Tikkanen, M. J. (2012). The effect of high-fiber rye bread enriched with nonesterified plant sterols on major serum lipids and apolipoproteins in normocholesterolemic individuals. Nutrition, Metabolism and Cardiovascular Diseases, 22(7), 575-582.

656. Sodhi, R.S. (2004). Indian Ice Cream Market: Amul Ice Cream Success Story, Indian Dairymen, 56(6), 66-67.

657. Sohail, M. N., Karim, A., Sarwar, M., & Alhasin, A. M. (2011). Onion (Allium cepa L.): an alternate medicine for Pakistani population.

International Journal of Pharmacology, 7(6), 736e744.

658. Soler-Vila, A., Couglan, S., Guiry, M.D. and Kraan, S. 2009. The red alga Porphyra dioica as a fish-feed ingredient for rainbow trout (Oncorhynchus mykiss): effects on growth, feed efficiency, and carcass composition. Journal of Applied Phycology, 21(5), 617-624.

659. Soni M., White LR., Kridawati A., Bandelow S, Hogervorst E. (2016). Phytoestrogen consumption and risk for cognitive decline and dementia: With consideration of thyroid status and other possible mediators. J Steroid Biochem Mol Biol. 160:67-77.

660. Sørensen, R., Kildal, E., Stepaniak, L., Pripp, A.H. and Sørhaug, T. 2004. Screening for peptides from fish and cheese inhibitory to prolyl endopeptidase. Nahrung/Food, 48(1):53–56

661. Souza, G. A., Ebaid, G. X., Seiva, F. R. F., Rocha, K. H. R., Galhardi, C. M., & Mani, F. (2011). N-acetylcysteine an allium plant compound improves high-sucrose diet-induced obesity and related effects. Evidence-based Complementary and Alternative Medicine, 643269e643274.

662. Spilburg CA, Goldberg AC, McGill JB, Stenson WF, Racette SB, Bateman J, McPherson TB, Ostlund RE Jr (2003). Fat-free foods supplemented with soy stanol-lecithin powder reduce cholesterol absorption and LDL cholesterol. Journal of the American Dietetics Association 103(5):577-81

663. Sridharan, E. (2004). The Growth and Sectoral Composition in India's Middle Class. India Review, 3(4), 405–428

664. Srinivasan, K., Sambaiah, K., & Chandrasekhara, N. (2004). Spices as beneficial hypolipidemic food adjuncts. A review. Food Review International, 20(2), 187e220.

665. Srivastava, KC., Bordia, A., and Verma, SK. (1995). Garlic (Allium sativum) for disease prevention. South Africa Journal of Sciience 91, p. 68-77.

666. Srivastava, R., Ahmed, H., Dixit Dharamveer, R., & Saraf, S. (2010). Crocus sativus L., a comprehensive review. Pharmacogn Review, 4(8), 200e208.

667. Stahl W, Sies H. Uptake of lycopene and its geometrical isomers is greater from heat-processed than from unprocessed tomato juice in humans. J Nutr. 1992;122(11):2161-2166.

668. Stanner SA, Hughes J, Kelly CNM, Buttriss J. A review of the epidemiological evidence for the "antioxidant hypothesis." Public Health Nutr. 2004;7(3):407-422.

669. Stappenbeck TS, Hooper LV, Gordon JI (2002) Developmental regulation of intestinal angiogenesis by indigenous microbes via Paneth cells. Proceedings of the National Academy of Sciences USA 99: 15451-15455

670. Stanton, C., Ross, R.P., Fitzgerald, G.F., & Van Sinderen, D. (2005). Fermented functional foods based on probiotics and their biogenic metabolites. Current Opinion in Biotechnology, 16, 198–203

671. Steinmetz, KA and Potter, JD (1991). Vegetables, fruit and cancer II. Mechanisms. Cancer Causes Control 2, p. 427-442.

672. Steinmetz, KA. and Potter, JD. (1991b). Vegetables, fruit and cancer I. Cancer Causes Control 2, p. 325-357.

673. Steinmetz, KA., Kushi, H., Bostick, RM., Folsom, AR, and Potter, JD. (1994). Vegetables, fruit, and colon cancer in the Iowa Women's Health Study. American Journal of Epidemiology 139, p. 1-15

674. Stilling RM, van de Wouw M, Clarke G, Stanton C, Dinan TG, et al. (2016) He neuropharmacology of butyrate: He bread and butter of the microbiota-gut-brain axis? Neurochem Int 99: 110-132.

675. Stoewsand GS. Bioactive organosulfur phytochemicals in Brassica oleracea vegetable - A review. Food Chem Toxicol. 1995; 33:537-43.

676. Street DA, Comstock GW, Salkeld RM, Schuep W, Klag MJ. Serum antioxidants and myocardial infarction. Are low levels of carotenoids and alpha-tocopherol risk factors for myocardial infarction? Circulation. 1994;90(3):1154-1161. (PubMed)

677. Stull, A. J., Cash, K. C., Johnson, W. D., Champagne, C. M., & Cefalu, W. T. (2010). Bioactives in blueberries improve insulin sensitivity in obese, insulin-resistant men and women. Journal of Nutrition, 140(10), 1764e1768.

678. Swinbanks D, O'Brien J (1993). Japan explores the boundary between food and medicine. Nature 364, 180.

679. Sugimoto, M., Arai, H., Tamura, Y., Murayama, T., Khaengkhan, P., Nishio, T., et al. (2009). Mulberry leaf ameliorates the expression

profile of adipocytokines by inhibiting oxidative stress in white adipose tissue in db/db mice. Atherosclerosis, 204(2), 388e394.

680. Sugiura M, Nakamura M, Ogawa K, Ikoma Y, Yano M. High serum carotenoids associated with lower risk for bone loss and osteoporosis in post-menopausal Japanese female subjects: prospective cohort study. PLoS One. 2012;7(12):e52643. (PubMed)

681. Sukandar, E. Y., Permana, H., Adnyana, I. K., Sigit, J. I., Ilyas, R. A., Hasimun, P., et al. (2010). Clin study of turmeric (Curcuma longa L.) and garlic (Allium sativum L.) extracts as antihyperglycemic and antihyperlipidemic agent in type-2 diabetes-dyslipidemia patients. International Journal of Pharmacology, 6(4), 438e445.

682. Suryanarayana, P., Saraswat, M., Mrudula, T., Krishna, T. P., Krishnaswamy, K., & Reddy, G. B. (2005). Curcumin and turmeric delay streptozotocin-induced diabetic cataract in rats. Investigative Ophthalmology and Visual Science, 46(6), 2092e2099.

683. Suzuki, R., Tanaka, M., Takanashi, M., Hussain, A., Yuan, B., Toyoda, H., et al. (2011). Anthocyanidins-enriched bilberry extracts inhibit 3T3-L1 adipocyte differentiation via the insulin pathway. Nutrition and Metabolism, 8, 14e19.

684. Suzuki, S. 1996. Studies on biological effects of water soluble homologous oligosaccharides of chitin and chitosans. Flavour and Fragrance Journal 15, 61-68.

685. Swinburn, B. A., Caterson, I., Seidell, J. C., & James, W. P. T. (2004). Diet, nutrition and the prevention of excess weight gain and obesity. Public Health and Nutrition, 7(1), 123e146.

686. Taing, M.-W., Pierson, J.-T., Hoang, V. L. T., Shaw, P. N., Dietzgen, R. G., Gidley, M. J., et al. (2012).

687. Mango fruit peel and flesh extracts affect adipogenesis in 3T3-L1 cells. Food Functions, 3(8), 828e836.

688. Takahashi, T., Tabuchi, T., Tamaki, Y., Kosaka, K., Takikawa, Y., & Satoh, T. (2009). Carnosic acid and carnosol inhibit adipocyte differentiation in mouse 3T3-L1 cells through induction of phase2 enzymes and activation of glutathione metabolism. Biochemictry Biophysics Research Communication, 382(3), 549e554.

689. Tan B. Palm carotenoids, tocopherols and tocotrienols. J Am Oil Chem Soc. 1989; 66:770-6.

690. Tan J, McKenzie C, Potamitis M, Horburn AN, Mackay CR, et al. (2014) The role of short-chain fatty acids in health and disease. Adv Immunol 121: 91-119.

691. Tan, W.S. and many co-authors. 2012. Red blood cell omega-3 fatty acid levels and markers of accelerated brain ageing. Neurology, 78(9), 658-664

692. Tauseef, S. M., Butt, M. S., & Anjum, F. M. (2009). Safety assessment of black cumin fixed and essential oil in normal Sprague dawley rats: serol and hematol indices. Food Chemistry and Toxicology, 47(11), 2768e2775.

693. Teikari JM, Laatikainen L, Virtamo J, et al. Six-year supplementation with alpha-tocopherol and β-carotene and age-related maculopathy. Acta Ophthalmol Scand. 1998;76(2):224-229. (PubMed)

694. Teixeira, L. G., Lages, P. C., Jascolka, T. L., Aguilar, E. C., Soares, F. L. P., Pereira, S. S., et al. (2012). White tea (Camellia sinensis) extract reduces oxidative stress and triacylglycerols in obese mice. Ciencia e Tecnolia de Alimentos, 32(4), 733e741.

695. Tesch BJ (2003). Herbs commonly used by women: an evidence-based review. American Journal of Obstetrics and Gynecology 188(5 Suppl):S44-55

696. The Alpha-Tocopherol, Beta Carotene Cancer Prevention Study Group. The effect of vitamin E and β-carotene on the incidence of lung cancer and other cancers in male smokers. N Engl J Med. 1994;330(15):1029-1035. (PubMed)

697. The Alpha-Tocopherol, Beta-Carotene Cancer Prevention Study Group. Effect of vitamin E and beta-carotene on the incidence of lung cancer and other cancers in male smokers. N Engl J Med. 1994;330(15):1029-1035.

698. The International Longevity Centre-UK. Older People and Functional Foods; 2010 [Accessed 22.02.17]. Available from: http://www.ilcuk.org.uk/files/pdf_pdf_152.pdf

699. Thomas PR, Earl R: Opportunities in the Nutrition and Food Sciences: Research Challenges and the Next Generation of Investigators. Edited by the Institute of Medicine's Food and Nutrition Board (IOM/NAS). Washington, DC: National Academies Press; 1994: 98-142.

700. Thomsen AB, Hansen HB, Christiansen C, Green H, Berger A (2004). Effect of free plant sterols in low-fat milk on serum lipid profile in hypercholesterolemic subjects. Eur J Clin Nutr

701. Thomsen AB, Hansen HB, Christiansen C, Green H, Berger A (2004). Effect of free plant sterols in low-fat milk on serum lipid profile in hypercholesterolemic subjects. Eur J Clin Nutr. 58(6):860-70

702. Thomsen AB, Hansen HB, Christiansen C, Green H, Berger A (2004). Effect of free plant sterols in low-fat milk on serum lipid profile in hypercholesterolemic subjects. European Journal of Clinical Nutrition 58(6):860-70.

703. Tijburg, LBM., Mattern, T., Folts, JD., Weisgerber, UM., and Katan, MB. (1997). Tea flavonoids and cardiovascular diseases: A review. Critical Reviews in Food Science and Nutrition 37, p. 771-785.

704. Tikkanen MJ, Hogstrom P, Tuomilehto J, Keinanen-Kiukaanniemi S, Sundvall J, Karppanen H. (2001). Effect of a diet based on low-fat foods enriched with nonesterified plant sterols and mineral nutrients on serum cholesterol. Am J Cardiol. 2001;88:1157–1162.

705. Trivedi, J.K., Sarren H., and Dhyani M. (2008). Rapid urbanization - Its impact on mental health: A South Asian perspective. Indian Journal of Psychiatry, 50(3), 161–165. 36

706. Trottier G., Bostrom P.J., Lawrentschuk N., Fleshner N.E. Nutraceuticals and prostate cancer prevention: a current review. Nat. Rev. Urol. 2010;7:21–30

707. Twain M (1897) Following the Equator: A Journey Around the World. Hartford: American Publishing Company.

708. Turnbaugh PJ, Ley RE, Hamady M, Fraser-Liggett CM, Knight R, Gordon JI. (2007) The human

709. Turnbaugh PJ, Ley RE, Mahowald MA, Magrini V, Mardis ER, et al. (2006) An obesity-associated gut microbiome with increased capacity for energy harvest. Nature 444: 1027-1031.

710. Udani, J. K., Singh, B. B., Singh, V. J., & Barrett, M. L. (2011). Effects of Acai (Euterpe oleracea Mart.) berry preparation on metabolic parameters in a healthy overweight population: a pilot study. Nutrition Journal, 10(1), 45e50.

711. United States Department of Agriculture (USDA) (2014): http://www.nal.usda.gov/fnic/foodcomp/search/. Accessed 19 Feb 2017

712. US Department of Agriculture, Agricultural Research Service. USDA Nutrient Database for Standard Reference; Release 28. 2015.

713. US Department of Agriculture, Agricultural Research Service. USDA Nutrient Database for Standard Reference; Release 28. 2015.

714. US Food and Drug Administration. Claims that can be made for conventional foods and dietary supplements. College Park, MD: FDA-CFSAN/Office of Nutritional Products, Labeling, and Dietary Supplements, 2003. www.cfsan.fda.gov/~dms/hclaims.html (Accessed 21.02.17).

715. USDA (2010). Functional Food Research in ARS. Agricultural Research, [online]. Available from: https://www.ars.usda.gov/ARSUserFiles/00000000/NPS/FinalFunctionalFoodsPDFReadVersion6-25-10.pdf [Accessed 24 January 2017].

716. Utsunomiya, H., Yamakawa, T., Kamei, J., Kadonosono, K., & Tanaka, S.-I. (2005). Anti-hyperglycemic effects of plum in a rat model of obesity and type 2 diabetes, Wistar fatty rat. Biomedical Research, 26(5), 193e200.

717. V. Liang, et al. (2005). Nutrient–gene interaction: metabolic genotype–phenotype relationship. Journal of Nutrition 135, pp. 3016S–3020S

718. Vadivel, V., Kunyanga, C. N., & Biesalski, H. K. (2012). Health benefits of nut consumption with special reference to body weight control. Nutrition, 28(12-Nov), 1089e1097.

719. Valencia, I., Ansorena, D. and Astiasaran, I. 2006. Nutritional and sensory properties of dry fermented sausages enriched with n-3 PUFAs. Meat Science, 72(4), 727-733

720. Valente, L.M.P., Bandarra, N.M., Figueiredo-Silva, A.C., Cordeiro, A.R., Simoes, R.M. and Nunes, M.L. 2007a. The influence of conjugated linoleic acid on growth, lipid composition and hepatic lipogenesis in juvenile sea bass (Dicentrarchus labrax). Aquaculture, 276(1-4), 225-235

721. Valente, L.M.P., Bandarra, N.M., Figueiredo-Silva, A.C., Rema, P., Vaz-Pires, P., Martins, S., Prates, J.A.M. and Nunes, M.L. 2007b.

Conjugated linoleic acid in diets for large-size rainbow trout
722. (Oncorhynchus mykiss): effects on growth chemical composition and sensory attributes. British Journal of Nutrition 97(2), 289-297

723. Van den Heuvel EG, Muys T, van Dokkum W, Schaafsma G. Oligofructose stimulates calcium absorption in adolescents. Am J Clin Nutr 1999;69:544–8.

724. Van der Made SM, Kelly ER, Kijlstra A, Plat J, Berendschot TT. Increased macular pigment optical density and visual acuity following consumption of a buttermilk drink containing lutein-enriched egg yolks: a randomized, double-blind, placebo-controlled trial. J Ophthalmol. 2016;2016:9035745. (PubMed)

725. Van Het Hof KH, West CE, Weststrate JA, Hautvast JG. Dietary factors that affect the bioavailability of carotenoids. J Nutr. 2000;130(3):503-506.

726. Van Kleef H, Van Trijp HCM, Luning P. Functional foods: health claim-food product compatibility and the impact of health claim framing on consumer evaluation. Appetite. 2005;44:299-308.

727. Van Loo J, Cummings J, Delzenne N, et al. Functional food properties of non-digestible oligosaccharides: a consensus report from the ENDO project (DGXII AIRII-CT94–1095). Br J Nutr 1999;81:121–32

728. Van Rensburg SJ, Daniels WM, van Zyl JM, Taljaard JJ. A comparative study of the effects of cholesterol, beta-sitosterol, beta-sitosterol glucoside, dehydroepiandrosterone sulphate and melatonin on in vitro lipid peroxidation. Metab Brain Dis. 2000;15:257–265

729. Van Vliet T. A double blind placebo controlled, parallel trial on the effect of oligofructose intake on serum lipids in male volunteers. Zeist, Netherlands: TNO, 1997.

730. Vanhanen HT, Miettinen TA: Effects of unsaturated and saturated dietary plant sterols on their serum contents. Clin Chim Acta. 1992, 205: 97-107.

731. Vanstone CA, Raeini-Sarjaz M, Parsons WE, Jones PJ: Unesterified plant sterols and stanols lower LDL-cholesterol concentrations equivalently in hypercholesterolemic persons. Am J Clin Nutr. 2002, 76: 1272-1278.

732. Vaquero, M. R., Yanez-Gascon, M.-J., Villalba, R., Larrosa, M., Fromentin, E., Ibarra, A., et al. (2012). Inhibition of gastric lipase as a mechanism for body weight and plasma lipids reduction in Zucker rats fed a rosemary extract rich in Carnosic acid. PLoS ONE, 7(6), e39773.

733. Vaskonen T, Mervaala E, Krogerus L, Karppanen H. (2002). Supplementation of plant sterols and minerals benefits obese zucker rats fed an atherogenic diet. J Nutr. 132:231–237.

734. Veeramachaneni S, Wang XD. Carotenoids and lung cancer prevention. Front Biosci (Schol Ed). 2009;1:258-274. (PubMed)

735. Vella, MN., Stratton, LM., Sheeshka, J.and Duncan, AM. (2014). Functional food awareness and perceptions in relation to information sources in older adults. Nutrition Journal 13: 44

736. Verbeke, W. (2005). Consumer acceptance of functional foods: Sociodemographic, cognitive and attitudinal determinants, Food Quality and Preference, 16, 45–57

737. Verbeke, W., Scholderer, J. and Lähteenmäki, L. Consumer appeal of nutrition and health claims in three existing product concepts. Appetite 2009, 52: 684-692

738. Vercruysse, L., van Camp, J. and Smagghe, G. (2005) ACE inhibitory peptides from enzymatic hydrolysates of animal muscle 33 protein: a review. Journal of Agricultural and Food Chemistry, 53(21), 2244-2245

739. Verhagen, H., Vos, E., Francl, S., Heinonen, M. and Van Loveren, H. Status of nutrition and health claims in Europe. Archives of Biochemistry and Biophysics 2010, 501: 6-15

740. Verhoeven, DTH., Goldbohm, RA., van Poppel, G., Verhagen, H., and van den Brandt, PA. (1996).

741. Epidemiological studies on brassica vegetables and cancer risk. Cancer Epidemiology, Biomarkers & Prevention. 5, p 733-748.

742. Virtanen, J.K., Siscovik, D.S. and Lemaitre, R.N. 2013. Circulating omega-3 polyunsaturated fatty acids and subclinical brain abnormalities on MRI in older adults: The Cardiovascular Health Study. Journal of the American Heart Association, doi 10.1161/JAHA.113.000305

743. Vishwanathan R, Kuchan MJ, Sen S, Johnson EJ. Lutein and preterm infants with decreased concentrations of brain carotenoids. J Pediatr Gastroenterol Nutr. 2014;59(5):659-665. (PubMed)

744. Vo, T.-S., & Kim, S.-K. (2013). Fucoidans as a natural bioactive ingredient for functional foods. Journal of Functional Foods, 5(1), 16e27.

745. Volpe R, Niittynen L, Korpela R, Sirtori C, Bucci A, Fraone N, Pazzucconi F (2001). Effects of yoghurt enriched with plant sterols on serum lipids in patients with moderate hypercholesterolaemia. Br J Nutr. 86(2):233-9.

746. Volpe R, Niittynen L, Korpela R, Sirtori C, Bucci A, Fraone N, Pazzucconi F (2001). Effects of yoghurt enriched with plant sterols on serum lipids in patients with moderate hypercholesterolaemia. British Journal of Nutrition 86(2):233-9

747. Voutilainen S, Nurmi T, Mursu J, Rissanen TH. Carotenoids and cardiovascular health. Am J Clin Nutr. 2006;83(6):1265-1271. (PubMed)

748. Vrieze A, Van Nood E, Holleman F, Salojärvi J, Kootte RS, et al. (2012) Transfer of intestinal microbiota from lean donors increases insulin sensitivity in individuals with metabolic syndrome. Gastroenterol 143: 913-916.

749. Walia, S., Singh, M., Kaur, C., Kumar, R., & Joshi, S. (2010). Antioxidant composition of red and orange cultivars of tomatoes (Solanum lycopersicon l): a comparative evaluation. Journal of Plant Biochemistry and Biotechnology, 19(1), 95e97.

750. Wang XD. Carotenoids. In: Ross CA, Caballero B, Cousins RJ, Tucker KL, Ziegler TR, eds. Modern Nutrition in Health and Disease. 11th ed: Lippincott Williams & Wilkins; 2014:427-439.

751. Wang Y, Chung SJ, McCullough ML, et al. Dietary carotenoids are associated with cardiovascular disease risk biomarkers mediated by serum carotenoid concentrations. J Nutr. 2014;144(7):1067-1074.

752. Wang Y, Cui R, Xiao Y, Fang J, Xu Q. Correction: effect of carotene and lycopene on the risk of prostate cancer: a systematic review and dose-response meta-analysis of observational studies. PLoS One. 2015;10(10):e0140415. (PubMed)

753. Wang Y, Cui R, Xiao Y, Fang J, Xu Q. Effect of carotene and

lycopene on the risk of prostate cancer: a systematic review and dose-response meta-analysis of observational studies. PLoS One. 2015;10(9):e0137427. (PubMed)

754. Ward, O.P. and Singh, A. 2005. Omega-3/6 fatty acids: alternative sources of production. Process Biochemistry, 40, 3627-3652.

755. Warshafsky, S., Kamer RS., and Sivak, SL. (1993). Effect of garlic on total serum cholesterol. A meta-analysis. Annals of Internal Medicine 119, p. 599-605.

756. Washida, K., Miyata, M., Koyama, T., Yazawa, K., & Nomoto, K. (2010). Suppressive effect of Yamato-mana (Brassica rapa L. Oleifera Group) constituent 3-butenyl glucosinolate (Gluconapin) on postprandial hypertriglyceridemia in mice. Bioscience, Biotechnology and Biochemistry, 74(6), 1286e1289.

757. Wasswa, J., Tang, J., Gu, X.H. and Juan, X.Q. 2007. Influence of the extent of enzymatic hydrolysis on the functional properties of protein hydrolysates from grass carp (Ctenopharyngodon idella) skin. Food Chemistry, 104(4), 1698-1704.

758. Weststrate JA & Meijer QW (1998) Plant sterol-enriched margarines and reduction of plasma total- and LDL-cholesterol concentrations in normocholesterolaemic and mildly hypercholesterolaemic subjects. European Journal of Clinical Nutrition 52, 334–343.

759. Weststrate JA, Meijer GW (1998). Plant sterol-enriched margarines and reduction of plasma total- and LDL-cholesterol concentrations in normocholesterolaemic and mildly hypercholesterolaemic subjects. European Journal of Clinical Nutrition 52(5):334-43

760. Wetterstrand KA, Deal C, et al. (2009). The NIH Human Microbiome Project. Genome Research. 19:2317–2323.

761. Witkin SS, Mendes-Soare H, Linhares IM, Jayaram A, Ledger WJ, et al. (2013) Influence of vaginal bacteria and D- and L-lactic acid isomers on vaginal extracellular matrix metalloproteinase inducer: implications for protection against upper genital tract. MBio 4: e00460.

762. Wolever TM, Spadafora P, Eshuis H (1991) Interaction between colonic acetate and propionate in humans. Am J Clin Nutr 53: 681-687.

763. Wolk, A., Larsson, S.C., Johansson, J.E. and Ekman, P. 2006. Longterm fatty fish consumption and renal cell wall carcinoma incidence in a population-based prospective cohort of women. The Journal of the American Medical Association, 296, 1371-1376.

764. World Cancer Research Fund International/American Institute for Cancer Research Continuous Update Project Report. Diet, nutrition, physical activity, and prostate cancer. 2014.

765. Wu HW, Wang Q, Ma TZ, Ren JJ. (2009). Comparative studies on the functional properties of various proteins concentrates preparations of peanut protein. Food Res Int. 42:343–348.

766. Wu HW, Wang Q, Zhou SM (2007) Research progress on peanut protein and its functional properties. China Oils and Fats 32(9):7–11

767. Wu J, Cho E, Willett WC, Sastry SM, Schaumberg DA. Intakes of lutein, zeaxanthin, and other carotenoids and age-related macular degeneration during 2 decades of prospective follow-up. JAMA Ophthalmol. 2015;133(12):1415-1424. (PubMed)

768. Xu, Q., & Si, L.-Y. (2012). Resveratrol role in cardiovascular and metabolic health and potential mechanisms of action. Nutr Research, 32(9), 648e658.

769. Yamada T, et al. (2010). A human gut microbial gene catalogue established by metagenomic sequencing.

770. Yang, CS. and Wang, Z-Y. (1993). Tea and cancer. Journal of the National Cancer Institute 85, p. 1038-1049.

771. Yang, L., Chen, J.-H., Lu, J., Wu, Q., Xu, T., Zhang, H., et al. (2012). Rice protein improves adiposity, body weight and reduces lipids level in rats through modification of triglyceride metabolism. Lipids Health and Disease, 11, 24e29.

772. Yatsunenko, T., Rey, F.E., Manary, M.J., Trehan, I., Dominguez-Bello, M.G., Contreras, M., Magris, M., Hidalgo, Gl, Galdassano, R.N., Anokhin, A.P., Heath, A.C., Warner, B., Reeder, J., Kuczynski, J., Caporaso, J.G., Lozupone, C.A., Lauber, C., Clemente, J.C., Knights, D., Knight, R. & Gordon, J.I. (2012). Human gut microbiome viewed across age and geography. Nature, 486, 222-228.

773. Yeum KJ, Russell RM (2002). Carotenoid bioavailability and bioconversion. Annu Rev Nutr. 22:483-504.

774. Yeum KJ, Russell RM, Krinsky NI, Aldini G. Biomarkers of antioxidant capacity in the hydrophilic and lipophilic compartments of human plasma. Arch Biochem Biophys. 2004;430(1):97-103

775. Yeum KJ, Russell RM. Carotenoid bioavailability and bioconversion. Annu Rev Nutr. 2002;22:483-504.

776. Yi W, Fischer J, Krewer G, Akoh CC (2005) Phenolic compounds from blueberries can inhibit colon cancer cell proliferation and induce apoptosis. J Agric Food Chem 53: 7320-7329.

777. You, W-C., Blot, WJ., Chang, Y-S., Ershow, AG., Yang, Z-T., An, Q. Henderson, B., Xu, G-W., Fraumeni, JF., and Wang, T-G. (1988). Diet and high risk of stomach cancer in Shandong, China. Cancer Research 48, p. 3518- 3523.

778. Young J: A perspective on functional foods. Food Science and Technology Today 1996, 10: 18-21. 47.

779. Young P, Cash BD: Probiotic use in irritable bowel syndrome. Curr Gastroenterol Rep. 2006, 8 (4): 321-326.

780. Yu JM, Ahmedna M, Goktepe I, Dai J. (2006). Peanut skin procyanidins : composition and antioxidant activities as affected by processing. J Food Compos Anal. 19:364–371

781. Yuda, N., Tanaka, M., Suzuki, M., Asano, Y., Ochi, H., & Iwatsuki, K. (2012). Polyphenols extracted from black tea (Camellia sinensis) residue by hot-compressed water and their inhibitory effect on pancreatic lipase in vitro. Journal of Food Science, 77(12), H254eH261.

782. Yusuf S, Dagenais G, Pogue J, Bosch J, Sleight P. Vitamin E supplementation and cardiovascular events

783. in high-risk patients: The Heart Outcomes Prevention Evaluation Study Investigators. N Engl J Med. 2000;342(3):154-160.

784. Zava DT., Dollbaum CM., Blen M. (1998). Estrogen and progestin bioactivity of foods, herbs, and spices. Proc Soc Exp Biol Med. 217(3):369-78

785. Zawistowski J (2008). Regulation of functional foods in selected Asian countries in the pacific rim. In

786. Nutraceutical and Functional Food Regulations in the United States and Around the World. 1st Edition. Edited by Bagchi D. USA: Elsevier Inc.

787. Zeisel S.H. Regulation of "Nutraceuticals." Science. 1999;285:185–186.

788. Zhang, N., Yamashita, Y. and Nozaki, Y. 2002. Effect of protein hydrolysate from Antarctic krill meat on the state of water and denaturation by dehydration of lizard fish myofibrils. Fisheries Science, 68(3), 672-679

789. Zhang, R., Xiao, W., Wang, X., Wu, X., & Tian, W. (2006). Novel inhibitors of fatty-acid synthase from green tea (Camellia sinensis Xihu Longjing) with high activity and a new reacting site. Biotechnical and Applied Biochemistry, 43(1), 1e7.

790. Zhao G, Liu Y, Zhao M, Ren J, Yang B. (2011) Enzymatic hydrolysis and their effects on conformational and functional properties of peanut protein isolate. Food Chemistry 127(4):1438–1443

791. Zhao, X.U., Jia, J. and Lin, Y. 1998. Taurine content in Chinese food and daily taurine intake of Chinese men. In: Lombardini, J.B., Schaffer, S.W. and Azuma, J. (eds). Advances in Experimental Biology and Medicine. Plenum Press, New York, USA, pp. 501-505.

792. Zielinska-Przyjemska, M., Olejnik, A., Dobrowolska-Zachwieja, A., & Grajek, W. (2009). In vitro effects of beetroot juice and chips on oxidative metabolism and apoptosis in neutrophils from obese individuals. Phytotherapy Research, 23(1), 49e55.

793. Zunino, S. J. (2009). Type 2 diabetes and glycemic response to grapes or grape products. Journal of Nutrition, 139(9), 1794Se1800S.

INDEX